BUILDING WIRELESS SENSOR NETWORKS

BUILDING WIRELESS SENSOR NETWORKS

THEORETICAL & PRACTICAL PERSPECTIVES

NANDINI MUKHERJEE

SARMISTHA NEOGY

SARBANI ROY

CRC Press
Taylor & Francis Group
Boca Raton London New York

CRC Press is an imprint of the
Taylor & Francis Group, an **informa** business

CRC Press
Taylor & Francis Group
6000 Broken Sound Parkway NW, Suite 300
Boca Raton, FL 33487-2742

First issued in paperback 2020

ISBN 13: 978-0-367-57535-9 (pbk)
ISBN 13: 978-1-4822-3006-2 (hbk)

Library of Congress Cataloging-in-Publication Data

Mukherjee, Nandini.
    Building wireless sensor networks : theoretical and practical perspectives / authors, Nandini Mukherjee, Sarmistha Neogy, and Sarbani Roy.
        pages cm
    Includes bibliographical references and index.
    ISBN 978-1-4822-3006-2 (alk. paper)
    1. Wireless sensor networks. I. Neogy, Sarmistha. II. Roy, Sarbani. III. Title.

TK7872.D48.M84 2016
681'.2--dc23   2015009941

Visit the Taylor & Francis Web site at
http://www.taylorandfrancis.com

and the CRC Press Web site at
http://www.crcpress.com

To my parents Binata Mukherjee and Prabhat Mukherjee
- Nandini Mukherjee

To my daughter Roshni and my husband Prasun Neogy
- Sarmistha Neogy

To my daughter Bhoomika, my husband Sujit Roy
and
my parents Irani Ghosh and Milan Ghosh
- Sarbani Roy

Contents

List of Figures

List of Tables

Preface

This is a wonderful time when people are experiencing the seamless convergence of a host of different happenings in the area of computer science and information technologies. Recent engineering advances have made this easier. A new generation of inexpensive devices (wireless sensor nodes) capable of collecting information with high-order accuracy has paved the way for developing networks that can be deployed for applications ranging from domestic to military. The technology for sensing along with communication and a little bit of processing includes electric and magnetic field sensors, seismic sensors, sensor arrays, location and navigation sensors, and infrared sensors, among others. Thus, wireless sensor networks have made their presence felt in all spheres of our life.

In the near future wireless sensor networking is expected to be the harbinger of a new generation of conveniences. With the advent of new design concepts and materials, performance and the prolonging of the lifetime of the network will improve. Wireless sensor networking still remains and will remain an exciting and emerging domain for researchers. At this interesting juncture, however, there is a very limited number of textbooks on wireless sensor networks. The books that are available are generally focused on some specific area of research. Thus, the coverage of many important topics of wireless sensor networks may not be adequate. Hence the books may not be able to serve the purpose of general students.

This book is intended to be a high-quality textbook for both undergraduate (prefinal and final years) and postgraduate levels. The book covers the important aspects of wireless sensor networks. It exploits the sensor network architecture, protocols, operating system, security and energy management. Additionally, it also provides working programming examples for students interested in experimentation. Since the book exploits the basic aspects, it will be a treasure for anyone interested and willing to move into the wireless sensor network enigma.

Acknowledgments

We would like to express our sincere thanks to the people who have helped us in various ways during the preparation of the manuscript of this book. We thank colleagues Ram Sarkar, Kaushik Ghosh, Zeenat Rehena, Suparna Biswas and Chandreyee Chowdhury for giving their time to read through the manuscript and providing insightful criticism. Thanks are due to our research scholars Suman Sankar Bhunia, Subrata Dutta, Sujoy Mistry, Atrayee Gupta, Tanmoy Maitra, Tathagata Das and Subhra Banerjee for lending their helping hands to the presentation of artwork and programming efforts. We thank all those responsible for publication of this book since its proposal to the publisher.

Authors

Nandini Mukherjee received her Ph.D. in computer science from the University of Manchester, UK, in 1999. She also received a Commonwealth Scholarship for her doctoral study in the UK. She completed a Master's in computer science and engineering from Jadavpur University, Kolkata, India in 1991, and a Bachelor of Engineering in computer science and technology from Bengal Engineering College, Sibpur, India in 1987.

Since 1992, Dr. Mukherjee has been a faculty member of the Department of Computer Science and Engineering at Jadavpur University. Currently, she is a professor in the department. She has also served as the director of the School of Mobile Computing and Communication at Jadavpur University for almost six years. Before joining Jadavpur University as a faculty member, Dr. Mukherjee also worked in the industry for approximately three years.

Dr. Mukherjee is an active researcher in her chosen field. Her research interests are in the areas of high performance parallel computing, grid and cloud computing and wireless sensor networks. She has published many research papers in internationally peer-reviewed journals and renowned international conferences. She also acted as a member of technical program committees and organizing committees and as a reviewer for many international conferences and renowned journals. In addition, Dr. Mukherjee acted as the lead investigator for many technical projects with social relevance. She is a senior member of IEEE and the IEEE Computer Society.

Sarmistha Neogy received her Ph.D. in engineering from Jadavpur University in 2006. She obtained her Master's in computer science and engineering and a Bachelor's in computer science and engineering from Jadavpur University, in 1989 and 1987, respectively.

Dr. Neogy served as a faculty member of University of Kalyani, West Bengal, India, and the University of Calcutta, before joining Jadavpur University in 2001. Presently, she is associate professor in the Department of Computer Science and Engineering.

Dr. Neogy's research interests are in the areas of fault tolerance in distributed systems, reliability and security in wireless and mobile systems and wireless sensor networks. She is a senior member of IEEE and the IEEE Computer Society. She has authored publications in international journals and proceedings of international conferences. She also acts as a member of technical program committees for international conferences and has reviewed for international journals. She has delivered tutorial lectures at international conferences.

Sarbani Roy received a Ph.D. in engineering from Jadavpur University in 2008, an M-Tech. in computer science and engineering, an M.Sc. in computer and information science and a B.Sc. (Hons) in computer science from University of Calcutta in 2002, 2000 and 1998, respectively. Since 2006, she has been a faculty member in the Department of Computer Science and Engineering, Jadavpur University.

Dr. Roy's research was focused on grid computing during her Ph.D. studies, but since 2009, she has been working in wireless sensor networks, with a focus on the design of energy efficient protocols. She has published research papers in internationally peer-reviewed journals and conference proceedings. Her research interests include distributed computing, wireless sensor networks, grid and cloud computing and social network analysis.

Dr. Roy received a Fulbright-Nehru Senior Research Fellowship from the United States-India Educational Foundation in 2013-2014. She has been involved in technical program committees and organizing committees for many international conferences and has also acted as a reviewer for many international conferences and journals. She is a member of IEEE and the IEEE Computer Society.

1

Introduction

Sensors have been in use for a very long time in traditional applications. These applications include touch-sensitive sensors on buttons of microwaves and elevators, motion detector sensors which turn lights on and off, smoke detector sensors, etc. With advances in electronics and communication technology sensors are now being used in many new applications which were not even considered only a decade back. Particularly, due to rapid development of wireless networks, it has been possible to deploy a network of sensors spread over a large geographical area to allow sensor devices communicating wirelessly to gather huge volume of data for use in several new applications which had not been envisioned earlier.

Our intent is to present to the readers the advancements in wireless sensor networks (WSNs), both in theoretical and practical perspectives and also enable them to write small applications in a WSN environment. Thus, with an overview of sensor and mote technologies and a discussion on issues and challenges of building WSNs in this chapter, the book will gradually focus on wireless communication protocol standards, routing and data aggreagation algorithms, localization techniques and algorithms, energy conservation and security issues. The last two chapters of the book concentrate on providing a practical guide to the readers for programming the sensor motes to develop small applications.

1.1   Sensors

Sensors are used to sense a wide range of parameters represented by different energy forms such as movement, electrical signals, thermal or magnetic energy, etc. Sensors are able to sense a physical change in some physical characteristic which changes in response to some excitation, for example, heat. The change in such a characteristic is converted to an electrical signal.

Any sensor produces a voltage (or signal) which is proportional to the change in the parameter measured. The type and amplitude of the output signal depend on the sensor type.

1

Active and Passive Sensors

Broadly, a sensor is classified as a *passive sensor* or an *active sensor*. Active sensors require an external power supply to operate. This external source of energy is called an excitation signal. Active sensors measure changes of their own properties in response to the external effect.

An example of an active sensor is a strain gauge which is used to measure strain on an object. Its electrical resistance can be measured by detecting variations in the current (or voltage) across it and relating these changes to the amount of strain or force applied, but this measurement requires a current to be passed through the gauge.

On the other hand, passive sensors do not require any additional energy sources. They are direct sensors which change their physical properties such as resistance and capacitance and generate electrical signals in response to an external stimulus. An example of a passive sensor is a photo-diode. When an external stimulus like light excites a photodiode, photons are abosrbed in the photodiode and current is generated.

Analog and Digital Sensors

Analog sensors produce continuous signal which is proportional to the parameter measured. Many physical parameters such as temperature, pressure, and displacement are analog quantities and they are measured as continuous signals.

Digital sensors produce discrete signals which are digital representations of the quantities of the parameters being measured. These discrete vaues are output as a single bit or a group of bits representing a quantity.

Properties of Sensors

A good sensor must obey the following rules:

- It should be sensitive to the measured property.

- It should be insensitive to any other property.

- It should not influence the measured property.

In an ideal situation, the output signal of a sensor is exactly proportional to the value of the measured parameter. The gain is then defined as the ratio between output signal and measured parameter. For example, if a sensor measures temperature and has a voltage output, the gain $[V/K]$ (V is voltage and K is temperature) is a constant with the unit.

An important consideration for a sensor is its area of coverage defined as the geographical region in the proximity of a sensor which is covered by it. A sensor can measure every change in the physical properties within that region.

TABLE 1.1
Commonly used sensors

Quantity Measured	Sensor
Light Level	Light Dependent Resistor (LDR)
	Photodiode
	Phototransistor
	Solar Cell
Temperature	Thermocouple
	Thermistor
	Thermostat
	Resistive Temperature Detector
Force/Pressure	Strain Gauge
	Pressure Switch
	Load Cell
Position	Potentiometer
	Encoder
	Reflective/Slotted Opto-switch
	LVDT
Speed	Tacho Generator
	Reflective/Slotted Optocoupler
	Doppler Effect Sensor
Sound	Carbon Microphone
	Piezoelectric Crystal

(Source: http://www.electronics-tutorials.ws)

1.2 Sensor Node Architecture

A sensor node, also known as a mote, is a building block in a wireless sensor network. In addition to sensing capabilities, a sensor node also wirelessly communicates with other nodes in the network to propagate information through the network. As we will see in the subsequent chapters, a sensor node must also be capable of performing some processing tasks. Thus, the major functionalities that a sensor node must perfom include sensing data from the environment, such as temperature, or motion, processing the data as required by the

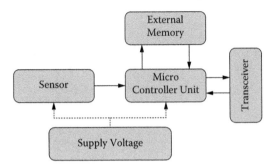

FIGURE 1.1
Sensor node architecture

application and communicating with other nodes in the network. Figure 1.1 depicts the architecture of a sensor node.

As depicted in the figure, a sensor node consists of a microcontroller, some amount of memory, a tranceiver and one or more sensors embedded in it. Sensor nodes are generally deployed in places where no external power sources are available. Therefore, sensor nodes are battery-powered and a power supply is integrated with it.

Sensing Subsystem

The sensing subsystem in a sensor node includes one or more sensors. For example, a node may be capable of sensing three parameters from the environment such as temperature, humidity and light if these three sensors are embedded in it. In the case of analog sensors, an analog-to-digital converter (ADC) is used to convert the analog output signal of a sensor into a digital signal.

Processing Subsystem

The processor subsystem interconnects all the other subsystems and some additional peripherals. Its main purpose is to execute instructions pertaining to sensing, communication and self-organization. This subsystem consists of a processor chip, a nonvolatile memory which stores program instructions, an active memory which temporarily stores the sensed data and sometimes processed data and an internal clock.

As a processing element, a mote (sensor node) often uses a microcontroller. A microcontroller contains a CPU core, a volatile memory (RAM) for data storage, a ROM, EPROM, EEPROM or flash memory, parallel I/O interfaces, discrete input and output bits, a clock generator, one or more internal analog-to-digital converters and serial communications interfaces.

Microcontrollers are of small size, low cost and their power consumption is low. Hence they are suitable for building computationally less intensive applications. However, microcontrollers are less powerful and less efficient in comparison with custom-made processors. The other options are *digital signal processors (DSPs), application-specific integrated circuits (ASICs)* and *field programmable gate arrays (FPGAs).*

DSPs process discrete signals with simple electronic circuits like adders, multipliers and delay circuits. Digital filters are used for reducing the noise effect and enhancing or modifying spectral characteristics. DSPs usually are based on Harvard architecture and are powerful and efficient. They can be used for applications where nodes are deployed in harsh physical settings and signal transmission may be affected by noise. However, DSPs are not suitable for tasks requiring periodic upgradation and modification.

An ASIC is actually an integrated circuit which can be customized for a specific application. Sometimes, a half-customized ASIC is built with logic cells that are available in the standard library. Whether an ASIC is fully customized or half-customized, the final logic structure is configured by the user. An ASIC can be optimized to meet the requirements of an application. However, its development cost is high and re-configuration is difficult. ASICs are used not to replace microcontrollers or DSPs but to complement them.

In comparison with ASICs, FPGAs are more complex in design and more flexible to program. FPGAs are programmed by modifying a packaged part. Programming is done with the support of circuit diagrams and hardware description languages, such as VHDL and Verilog. Although, FPGAs are complex and undergo an expensive design and realization process, there are some advantages of using them. FPGAs have higher bandwidth compared to DSPs, support parallel processing, can work with floating point representations and provide greater flexibility of control.

Communication Subsystem

In a wireless sensor network, fast and energy-efficient data transfer between the sensor nodes is important. However, the sensor node sizes are made small, so that they can be deployed on a large scale over a large geographical area. The size of sensor nodes puts restrictions on system buses and parallel transmission cannot be supported. Usually a high speed, full duplex, synchronous serial bus is used for communication.

A transceiver which has both a transmitter and a receiver that share common circuitry is used for communication. Data is transmitted as electromagnetic signals at radio frequencies. Radio transceivers transmit a bit or a byte stream as a radio wave. Transceivers can be put into different operational states: typically transmit, receive, idle (ready to receive, but not doing so) and sleep (significant parts of the transceiver are switched off). In sleep mode, a receiver is not able to receive immediately.

Different communication protocols are used for sensor networks. These include IEEE 802.11, IEEE 802.15.4, ZigBee and Bluetooth. Recently, a new standard, called 6LoWPAN has been proposed to enable sensor devices to communicate over IPv6. The communication protocols are discussed in detail in Chapter 2.

Energy Supply

The sensor nodes are built for deployment in regions where power supply from mains is not available. Therefore, batteries are used as main sources of energy for sensor nodes. Often non-rechargeable primary batteries are used. Sometimes rechargeable secondary batteries are used in combination with some form of energy harvesting. Energy can be scavenged from the environment, using solar cells, air or liquid flow.

Reducing energy consumption is important for battery-powered sensor nodes. In order to save power, sensor nodes are used along with multiple power consumption modes. If they have nothing to do, nodes are switched to power safe mode. Typical modes for different components of a sensor node are as follows. A controller can be in active, idle or sleep mode. The key to *low duty cycle operation* is that a controller must *sleep* most of the time and when it *wakes up* it should quickly start processing. It must also minimize its work in its *active* mode and return to sleep mode. A controller in *sleep* mode can also minimize sleep current through isolating and shutting down individual circuits. ADC conversions, DMA transfers, and bus operations are performed when the microcontroller core is stopped.

As mentioned earlier, transmitters and receivers can be put in active, idle or sleep states. In sleep state, the transmitter is turned off or the receiver is turned off or both are turned off. Although, energy consumption can be reduced by keeping receivers in sleep mode when they are not in use, the recovery time and startup energy to transfer from sleep state to wakeup state can be significant. The other issue is when to wake a receiver. Some of the MAC layer protocols to handle these issues are disucssed in Chapter 4.

1.3   Sensor Network Architecture

In a wireless sensor network, spatially distributed tiny devices are deployed to monitor physical or environmental conditions, such as temperature, sound, vibration, pressure, motion or pollutants. Each node or device has sensing, processing and communication capabilities and they cooperatively pass the gathered data through the network to a main location, where the base station is located or a gateway is connected to an external network such as Internet.

Participatory Roles of Sensor Nodes

In a wireless sensor network, nodes can act as *source* nodes, that is, as entities which sense data from the environment. One or more nodes act as sink nodes

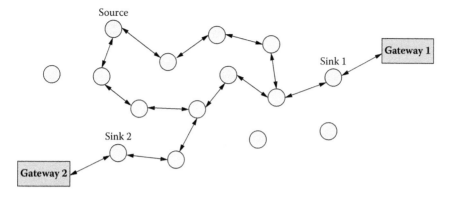

FIGURE 1.2 (SEE COLOR INSERT)
Multi-hop sensor network

where information is passed on. Sinks may belong to the sensor network itself or they may be external entities, such as laptops or personal digital assistants (PDAs), but directly connected to the network. The main difference is that an external sink comes and goes and often moves around, whereas when a sink is part of a network, it is generally static or its mobility is restricted. A sink may also be part of an external network (e.g., Internet), which is somehow connected to the sensor network. A sink can act as a base station or there can be a separate high-end device acting as a base station or a gateway connected to the Internet.

A wireless sensor network is essentially an ad hoc network, like the *mobile ad hoc NETwork* or MANET. Due to limited transmission range of the sensor nodes, power constraints, physical obstacles in the geographical region and path losses, any source in the network cannot directly communicate with the sink. Therefore, a *store-and-forward* multi-hop network is used by routing packets through intermediate nodes. Intermediate nodes forward packets to the destination, i.e., the sink.

Deployment of Sensor Nodes

A wireless sensor network can be deployed as a well-planned network and any of the common topologies like star, mesh, etc. can be used. This is known as *regular deployment* and is used in applications like preventive maintenance. Regular deployment may not necessarily have a geometric structure, but this is often a convenient assumption.

Well planned regular deployment is generally not possible for many applications. For example, in case of battlefield monitoring, sensor nodes may be dropped from an aircraft. In such situations, *random deployment* follows. For experimental purposes, uniform random distribution for nodes over finite area is assumed.

Sensor nodes can also be mobile. In a mobile sensor network, the nodes move to compensate for various shortcomings of deployment. Sensor nodes can be passively moved around by external forces like wind and water. Mobile nodes can also actively seek out interesting areas where certain types of events occur.

1.4   Mote Technology

Motes are small, inexpensive, low-power devices that can automatically form ad hoc wireless communication networks among themselves when deployed in an indoor or outdoor environment. A mote typically has several components: a radio transceiver with an internal antenna or connection to an external antenna, a microcontroller, an electronic circuit for interfacing with the sensors and an energy source, usually a battery or an embedded form of energy harvesting, such as solar panel. When sensors are connected to the I/O interfaces of a mote, the mote becomes a cost-effective platform for distributed sensing applications. It runs embedded programs to collect, store and transmit the measurements collected by sensors integrated with it.

A DARPA-sponsored "smart dust" project was initiated in late 1990s. The project envisioned building miniature MEMS-based devices of the order of 1 mm in size with an integrated solar cell, thick film battery, analog sensor(s), a small processor, and optical transceiver.

At the same time, a number of larger macro mote devices have also been produced using inexpensive commercial components available in the market. The term *mote* was first coined at University of California, Berkeley [1]. The basic hardware design principle for these motes is to integrate sensors, computation and communication in a single unit, built on a basic board with radio, processor and memory and these boards are sandwiched in layers. Generally, open-source hardware and software concepts are used. Also modular design has been adopted for fast development.

Some popular motes [2] are: Mica, Mica2, Mica2Dot, MicaZ, Telos, TelosB, IRIS and Imote. Some commonly available motes and their capabilities are discussed in this section.

Rene

The first commercial generation of this platform was the Rene mote. The Berkley Rene motes were developed in 1999 by Crossbow Technologies. Rene was based on AT90LS8535 processor with 8 KB of program memory and 0.5 KB of RAM. It used a radio with data rate of 10 Kbps and OOK encoding.

Mica Motes

Next, Berkeley and the collaborating researchers devised a second-generation platform named Mica. Mica motes were developed as a series of sensor devices embedded with processor, radio and sensor circuit on thin and small circuit boards. The hardware design consisted of a low-power radio and processor board (known as a mote processor/radio or MPR board) and one or more sensor boards (known as a mote sensor or MTS board). The MPR board included a processor Atmel ATmega103 which is a 4-Mhz, 8-bit CPU with 128 KB instruction memory and 4 KB SRAM and EEPROM. It also included radio operating at 50 Kbps with ASK modulation with a communication range up to 300 ft, 4 Mb flash with serial peripheral interface (SPI), 51-pin connector, an A/D converter and battery.

Mica2, the crossbow third generation mote, brought design changes in Mica. Processor Atmel ATmega128L offered a stand-alone boot loader. These motes could work with TinyOS version 1.0 (TinyOS is an open source operating system designed for low-powered wireless devices, typically sensor motes). The radio range in Mica2 increased to 500 to 1000 ft. The radio operates at 38.4 Kbps with FSK encoding. Communication efficiency also increases due to FM modulation, built-in Manchester encoding, software programmable frequencies and better noise immunity. Mica2 also includes a 512-Kb serial flash.

Mica2DOT was developed as Crossbow's third-generation mote with similar feature set to Mica2. Its size reduced to 25 mm diameter and 6 mm height and I/O capabilities degraded by reducing the number of connector pins to only 18 instead of 51 as in Mica2. The mote was integrated with temperature and battery voltage sensors.

Telos Platforms

Crossbow offers the TelosB mote (TPR2400) as an open source platform designed to enable cutting-edge experimentation for research communities. TelosB uses a 16-bit RISC processor TI MSP430 with 8-MHz processor speed, 10 KB RAM, 48-KB program space and 512/1024-KB flash memory. TelosB uses a 802.15.4-compliant CC2420 RF transceiver operating at 2.4 Ghz and with 250 Kbps data rate. The radio range is 100 m.

TelosB motes (TPR2420) are integrated with an optional sensor suite that includes light, temperature and humidity sensors.

IRIS Motes

IRIS motes are designed with Atmel ATmega1281 processors offering 128 Kb program flash memory, 512 Kb flash, 8 Kb RAM and 4 Kb EEPROM integrated with UART serial communication and a 10-bit analog-to-digital converter. IRIS motes include IEEE 802.15.4-compliant transceivers which operate at 2.4 to 2.48 GHz with direct sequence spread spectrum and offer 250 kbps data rate.

FIGURE 1.3 (SEE COLOR INSERT)
Sensor motes: (a) Rene, (b) Mica2, (c) Mica2DOT

Sensor boards and data acquisition cards are connected to motes to provide direct sensing capabilities. MEMSIC offers a variety of sensor and data acquisition boards for the IRIS motes. All of these boards connect to the IRIS via the standard 51-pin expansion connector. MTS400, MTS420 and MTS310 are popular sensor boards for environmental applications to measure humidity, temperature, pressure, light, etc. MDA300 is a mote data acquisition board which includes external environmental probes for sensing humidity, soil moisture, PAR light, wind speed and direction. It has 8 analog inputs, 8 digital input/outputs, 2 relay channels and selectable sensor excitation of 2.5V, 3V and 5V.

Multiple interfaces, including Ethernet, Wifi, USB and serial are supported by gateway boards to provide a base station for connecting an IRIS or Mica sensor network to a PC or workstation. Any Mica or IRIS mote can function as a base station when connected to the relevant interface board.

These motes can be programmed using programming boards. MIB520 is a USB program board that provides interface between a mote and a PC. It has a parallel port, a more commonly used serial port and ethernet connectivity.

FIGURE 1.4 (SEE COLOR INSERT)
Sensor motes: (a) TelosB, (b) IRIS

1.5 Comparison of MANET and WSN

A wireless sensor network is similar to a *mobile ad hoc NETwork* or MANET. Here also no pre-deployed infrastructure is required. The nodes are organized in the form of flexible mesh architectures which can dynamically adapt to support introduction of new nodes or expand to cover a larger geographic region. As long as there is sufficient density, a single network of nodes can grow to cover an unlimited area. Unlike cell phone systems that deny services when there are many phones active in a small area, the interconnection of a wireless sensor network only grows stronger when nodes are added. The system can automatically adapt to compensate for node failures as well. In traditional wireless systems, nodes directly communicate with the nearest high-power control tower or base station. However, in cases of WSN and MANET, nodes only interact with their local peers and forward data packets through multi-hop networks. Thus, like MANET, WSN is also self-organizing, energy efficient and generally a wireless multi-hop network. However, there are many differences which are discussed below.

MANETs are built with powerful (also expensive) equipment, with higher data rates and more resources. On the other hand, sensor nodes are tiny devices which are deployed to create a mesh network. The nodes have limited processing power and are energy constrained.

The main purpose of MANET is to interact with humans. Thus, MANETs are often used for human-in-loop applications. In contrast, in a WSN, nodes do not interact with humans; rather they interact with the environment, which is absent in MANET. MANETs are also comparably uniform.

Sensor networks are generally application-specific. Therefore, the QoS requirement is different for different applications. A sensor network can be deployed on a large scale compared to MANETs and they are deployed in places where maintenance can be difficult.

In WSNs, as long as there is sufficient density to construct a multi-hop network, individual nodes are dispensable. This is not the case for MANET. We will see in the subsequent chapters that WSNs are data-centric and in-network processing is a major requirement. In a WSN, accomplishing a task is important, not which node is performing the task. On the other hand, like traditional networks, MANETs are id-centric and each node in a MANET is addressed with an identifier and each node is given a specific task to accomplish.

1.6 Requirements of a WSN

Traditional networks transmit bits or signals through the network. In contrast with traditional networks, wireless sensor networks are more application-specific. Thus, they need to look for answers, instead of just moving bits through the network. Therefore, *quality of service* requirements are different for these networks. The major requirement is that the right answer must be provided at the right time.

Fault tolerance is another important requirement. A WSN must be robust against node failure. Usually, in a harsh environment, nodes may run out of energy or get destroyed physically. Thus, fault-tolerant and energy-efficient algorithms should be implemented to increase the lifetime of a network. It must be noted that lifetimes of individual nodes are relatively unimportant, because death of an individual node does not affect the activity of a network as long as the neighboring nodes can accomplish the assigned task. Lifetime of the entire network is therefore important, though definition of lifetime depends on application. In many applications, "death of the first node" is considered as the network lifetime, whereas in certain cases a network can remain active until the "death of the last node." There can be other definitions of lifetime as well.

Flexibility and universality of a network are the other requirements. Devices or nodes must be used in a wide range of application scenarios. Auto-configuration is also necessary.

An important requirement of a wireless sensor network is that nodes in the network must collaborate towards a joint goal. Often, data are gathered from all nodes and all these data collectively provide some information to the end user. For example, in order to compute the average temperature in a region, all nodes must cooperate to find the answer. Similarly, to detect a boundary of a disaster region, nodes must cooperatively run an edge detection algorithm. Instead of transmitting huge volume of data towards the sink and then using them for finding a solution, data should be processed within the network. The general tendency is to perform local computation as much as possible (on nodes or among nearby neighbors) instead of forwarding every piece of data towards the base station.

Pre-processing of data within the network also greatly improves its energy efficiency. However, due to low computational capabilities of individual nodes, there is a chance of loss of accuracy. A tradeoff must be made between energy consumption and accuracy.

1.7   Challenges for a WSN

The major challenge for a WSN is to cope with the harsh resource constraints placed on the individual devices. Because sensor devices should be produced in vast quantities and must be small and inexpensive, all resources including processor, memory and energy are limited on a sensor node. Only a small memory is used to implement complex, distributed, ad hoc networking protocols. The memory can store only a small amount of data. Processors also have low speed and low processing power.

As the physical size of a sensor node decreases, the node becomes more energy constrained. Underlying energy constraints also lead to further computational and storage limitations. A node must use low duty cycle operation to extend its lifetime. The radio capacity is also decreased, though raw channel capacity is much greater.

1.8 WSN Applications

Wireless sensor networks can be deployed and used in various application areas. These areas include environment monitoring, biodiversity mapping, healthcare monitoring, forest fire detection, landslide detection and intrusion detection.

For example, sensor nodes may be dropped from an aircraft over a wildfire. Each node measures temperature and they collaboratively derive a temperature map.

Sensor nodes can be used to observe wildlife and create a biodiversity map.

Intelligent buildings (or bridges) can be constructed where sensor nodes are used to reduce energy wastage by proper humidity, ventilation, and air conditioning (HVAC) control. This application requires measurements of room occupancy, temperature, air flow, etc.

Goods (parcels, containers) can be equipped with sensor nodes so that their whereabouts can be tracked.

Another area for sensor network application is precision agriculture. Decisions on fertilizers, pesticides and irrigation are made only where needed.

WSNs can be used for controlling leakages in chemical plants and for machine surveillance and preventive maintenance. In certain sensing and control functions, it is required to deploy a wireless sensor network when it is not possible to deploy a wired network or a human cannot access the area.

Medicine and health care applications are becoming the focus in wireless sensor network research. Sensor networks can be used for post-operative or intensive care, long-term surveillance of chronically ill or elderly patients and telemedicines.

WSNs can also provide better traffic control by obtaining finer-grained information about traffic conditions. Intelligent roads can be created by deploying sensors or cars can be integrated with sensor nodes.

Internet of Things

The "Internet of Things" (IoT) describes an environment where physical objects can be accessed through the Internet any time from anywhere. Wireless sensor networks form basic ingredients for developing the IoT applications. A world is envisioned where, like the five sensing organs of a human body, WSNs will perform the sensing tasks. As the brain processes all data sensed by human organs, a huge volume of data collected by the WSNs will be processed on large computing servers (probably in a cloud computing environment) connected to the Internet and will be used for analysis and reasoning over the Internet and possibly for remote actions.

A number of projects have been started for real-life applications of WSNs and IoT. A number of architectural, protocol and other issues require further research. Handling a large volume of data is another issue which is being considered by the "big data" researchers.

1.9 Chapter Notes

For developing prototype applications based on wireless sensor networks, mote technology has been introduced by the research community. Many of these motes have been designed and developed with open source hardware and software. Detailed design and other information is available on the Internet.

Some information on Mica motes is also available [6, 4, 3]. Design of Telos motes is described in References [5, 7].

Bibliography

[1] Website: http://smote.cs.berkeley.edu/motescope/.

[2] Website: http://www.memsic.com/wireless-sensor-networks/.

[3] P. Ballal and F. Lewis. Introduction to Crossbow Mica2 sensors.

[4] Crossbow. Mica2: Document part number: 6020-0042-04. Website: www.xbow.com.

[5] J. Polastre, R. Szewczyk, and D. Culler. Telos: Enabling ultra-low power wireless research. In *Proceedings of the 4th International Symposium on Information Processing in Sensor Networks*, IPSN '05, Piscataway, NJ, USA, 2005. IEEE Press.

[6] D. Rossi. Sensors as hardware: Motes evolution.

[7] Willow Technologies. TelosB mote platform. Website: www.willow.co.uk/TelosB_Datasheet.pdf.

2

Wireless Sensor Networks Architecture

2.1 Introduction

Advances in wireless communication and electronics during the last couple of decades have made it possible to deploy multi-hop wireless sensor networks for various applications. These wireless sensor network applications require that the network be ad hoc, self-organising and self-configurable in nature. The communication protocol, network architecture must be standardized to fulfill the above requirements. A common sensor network architecture has many benefits in terms of interoperability and code reuse. Researchers have been engaged for a long time to propose an acceptable sensor network architecture to support all types of sensor devices and their various applications. Nevertheless, in spite of the existence of a number of proposals, the major challenge that hindered the standardization of network architecture and communication protocols is the increasing heterogeneity in sensor devices and their applications.

In this chapter, discussion focuses on a proposed network architecture and network protocol stack for wireless sensor networks. Communication standards which have been implemented and are used in wireless sensor network applications are also discussed.

2.2 Network Protocol Stack

Akyildiz et al [1] presents a protocol stack which is depicted in Figure 2.1. The stack consists of five layers: a physical layer, data link layer, network layer, transport layer and application layer. It also includes a power management plane, mobility management plane, and task management plane.

The physical layer implements modulation, transmission, and receiving techniques, which must be lightweight and robust. The medium access control (MAC) layer protocol addresses the issues like energy conservation and collision avoidance with the broadcasts from neighboring nodes. Network layer takes care of routing the data in the multi-hop network. Transport layer helps to maintain the flow of data if the sensor network application requires it.

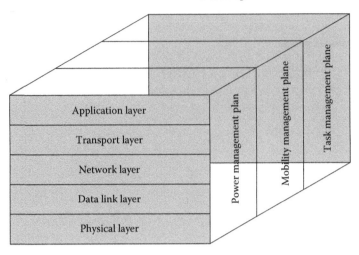

FIGURE 2.1
Layered protocol stack [1]

Depending on the sensing tasks, different types of application software can be built and used on the application layer.

In addition, the power, mobility, and task management planes monitor the power, movement and task distribution among the sensor nodes. These planes help the sensor nodes coordinate the sensing task and lower overall power consumption.

Physical Layer

In a multi-hop sensor network, transmission medium is wireless and links are formed by radio, infrared, or optical media. Currently available sensor nodes are based on radio frequency (RF) circuit design. Thus, the physical layer primarily deals with issues releated to communication based on radio frequency-based links.

Reducing energy consumption is one of the major issues in wireless sensor networks. In general, the minimum output power required to transmit a signal over a distance d is proportional to d^n, where $2 <= n < 4$. In case of sensor devices, antennas are of low height and channels are close to the ground level. Therefore, exponent n is close to four in sensor network communication [15]. Measurements carried out in [18] indicate that the power starts to drop off with higher exponents at smaller distances for low antenna heights. Another requirement is to overcome shadowing and path loss effects. Energy-efficient hardware design for overcoming the above problems is an open research problem. Similarly, low-power modulation schemes need to be developed for sensor networks. Energy-efficient physical layer solutions have been investigated by many researchers [17, 19, 8]. Nevertheless, many of the problems are yet to be explored.

Data Link Layer

The data link layer is responsible for multiplexing data streams, data frame detection, medium access and error control. In the data link layer of an infrastructureless network, as in the case of WSNs, the MAC schemes require establishing communication links for hop-by-hop data transfer and provide self-organizing ability to the network. Furthermore, the MAC layer must also provide support for efficient sharing of communication resources between sensor nodes. Energy conservation is another important issue that needs to be handled at the MAC layer . Error control schemes should also be lightweight and suitable for low-powered devices with limited computational ability. MAC layer protocols will be discussed in detail in Chapter 4.

Network Layer

Sensor networks are infrastructureless and multi-hop. Each sensor node is able to sense data as a source node and also forward the data sensed by it or by other nodes towards the sink. Thus, it is the responsibility of the network layer to route the data towards the sink. However, traditional ad hoc network protocols are not suitable for sensor networks because unlike traditional ad hoc networks, these networks are data-centric. Moreover, power efficiency is always an important consideration for sensor networks. Energy efficient routing has been investigated by many researchers and many routing algorithms have been proposed. Routing protocols will be discussed in detail in Chapter 3 and Chapter 4.

Transport Layer

The transport layer is needed when the system is planned to be accessed through the Internet or other external networks.

However, in a wireless sensor network, most research work and implementations targetted protocols at the data link and network layers. 6LoWPAN, a communication standard proposed for Internet-based wireless sensor networks, adopts upper layers of an IP-based network by inserting an adaptation layer between the IPv6 network layer and the IEEE 802.15.4 MAC layer.

2.3 Communication Standards

In this section, already established communication standards for wireless LAN and wireless PAN are discussed briefly. These standards are relevant for establishing wireless sensor networks. The objective is to give an overview of each standard and highlight its important characteristics.

2.3.1   IEEE 802.11

Wireless local area networks (WLANs) generally use the IEEE Standard 802.11, commonly known as Wi-Fi. With operating speeds comparable to wired networks and due to several additional advantages like flexibility and performance, Wi-Fi is now widespread and has become the de facto standard for WLANs.

2.3.1.1   General Description

The basic building block of an IEEE 802.11 [3] WLAN is a basic service set or BSS. Within a BSS, there are stations. A station is any device that contains functionalities of 802.11 physical and MAC layers. The 802.11 standard supports the two topologies:

- Independent basic service set (IBSS) networks - An IBSS is a stand-alone BSS (infrastructureless or without a backbone) and consists of at least two wireless stations. This type is often referred to as an ad hoc network.

- Extended service set (ESS) networks - An ESS network supports multiple cells interconnected by access points and a distribution system, such as ethernet. Within an ESS, a station can move within a local BSS; it can also move from one BSS to another BSS within the same ESS.

BSSs can partially overlap or may be physically disjointed. BSSs can also be physically collocated when redundancy is required for higher performing network.

2.3.1.2   MAC Layer

The IEEE 802.11 MAC layer supports a MAC sublayer which defines the access mechanisms, fragmentation, and encryption. There is also a MAC management layer for roaming, MIB (management information base) and power management.

Within 802.11 there are two options for the MAC layer : (i) a point coordination function (PCF) which is a centralised control scheme and (ii) a distributed coordination function (DCF) which is a contention-based approach.

In the PCF mode, wireless access points become point coordinators and coordinate the communication within the network. Access points periodically send beacon frames to communicate network management and identification which is specific to that WLAN. Between the sending of these frames, PCF splits the time frame into a contention-free period (CFP) and a contention period. The contention-free period is the time period when the stations (STAs) get rights to transmit solely by a point coordinator (PC), allowing frame exchanges to occur between members of the basic service set (BSS) without contention for the wireless medium. If PCF is enabled on the remote station, it can transmit data during the contention-free polling periods.

Thus, at the beginning of a contention-free period, all stations in the basic service area of the access point are informed that the medium is idle for a certain duration known as a point coordinate inter-frame space (PIFS). During the CFP, the access point polls each station in its polling list. If a station receives a polling frame from the access point, it can respond to the access point after a short inter-frame space (SIFS) period with an acknowledgement frame. If the access point or the station is ready to send data, the polling frame (from the access point) and the acknowledge frame (from the responding station) may be sent along with data packets. A station can respond to the access point with a DATA frame or a NULL frame. The access point continues to poll each station until it reaches the maximum duration of the CFP. The access point can also terminate the CFP.

Beacon frames contain information about the maximum duration of the CFP, beacon interval, and BSS identifier. All stations in the BSS set their NAVs and do not initiate any transmission of packets during the CFP after receiving a beacon. During the contention period, the MAC layer operates under DCF scheme.

The other scheme, DCF, uses CSMA-CA (carrier sense multiple access with collision avoidance). When this technique is used, the MAC layer of a node wanting to make a transmission listens for a clear channel and when the channel appears to be idle, it waits for a pre-defined period (known as distributed inter-frame space (DIFS)) and starts decreasing the backoff counter in a step-wise manner. As soon as the back-off timer expires, the station transmits its packet. The receiver acknowledges the transmitted packet after a short inter-frame space (SIFS) if the packet is received correctly. If a clear channel is not available (other nodes are transmitting) or if the node does not receive an acknowledgement after transmission of a packet, it backs off a random amount of time, and then listens for a clear channel and retransmits the data [16].

2.3.1.3 Physical Layer

IEEE 802.11 has been generalized so that it can operate independently of license type, band, and country of operation. Thus, these Wi-Fi standards operate within the industrial, scientific and medical (ISM) frequency bands. These are shared by a variety of other users, but no license is required for operation within these frequencies. Hence, these bands are ideal for a general system and for widespread use. These standards define the access mechanisms, handling fragmentation, encryption, power management, synchronization,and roaming at the MAC layer level. At the physical layer level, are a PLCP sublayer and a PMD sublayer. The 802.11 standards use clear channel assessment signal (carrier sense) as a *physical layer convergence protocol (PLCP)* at the PLCP sublayer level. The *physical medium dependent sublayer* or PMD sub-

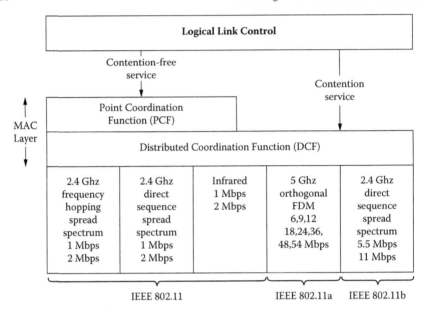

FIGURE 2.2
IEEE 802.11 protocol architecture

layer defines modulation and coding. The *physical layer management* component interacts with these two layers and control channel selection and manages
the MIB (management information base). Station management mechanisms
are implemented by a *station management layer*.

The physical layer of IEEE 802.11 supports five versions. Among these
there is one infrared version. Other four are based on two radio frequencies,
2.4 GHz and 5 GHz, and data rates of 1 or 2 Mbps. Either frequency hopping
spread spectrum or direct sequence spread spectrum modulations are supported. With DSSS modulation, higher data rates and ranges can be achieved,
though the cost and power consumption are higher in comparison with FHSS.
Figure 2.2 provides an overview of 802.11 protocol architecture.

2.3.1.4 Standards

IEEE 802.11 defines a variety of standards, each named with a letter suffix.
These standards are specifically defined to address security aspects, quality
of service aspects and other issues. The standards defined under IEEE 802.11
are as follows:

- 802.11a - Wireless network bearer operating in the 5 GHz ISM band with
  data rate up to 54 Mbps

- 802.11b - Wireless network bearer operating in the 2.4 GHz ISM band
  with data rates up to 11 Mbps

- 802.11e - Quality of service and prioritisation

- 802.11f - Handover

- 802.11g - Wireless network bearer operating in 2.4 GHz ISM band with data rates up to 54 Mbps

- 802.11h - Power control

- 802.11i - Authentication and encryption

- 802.11j - Interworking

- 802.11k - Measurement reporting

- 802.11n - Wireless network bearer operating in the 2.4 and 5 GHz ISM bands with data rates up to 600 Mbps

- 802.11s - Mesh networking

- 802.11ac - Wireless network bearer operating below 6GHz to provide data rates of at least 1Gbps per second for multi-station operation and 500 Mbps on a single link

- 802.11ad - Wireless network bearer providing very high throughput at frequencies up to 60GHz

- 802.11af - Wi-Fi in TV spectrum white spaces (often called white-Fi)

Among these standards, 802.11a, 802.11b, 802.11g and 802.11n are widely used for real-life applications. The 802.11n standard is the latest one which provides raw data rates up to 600 Mbps.

The first accepted 802.11 WLAN standard was 802.11b. This standard used frequencies in the 2.4 GHz ISM frequency band and offered raw data rates of 11 Mbps. The modulation scheme used by this standard is complementary code keying (CCK). CCK is a slight variation on CDMA (code division multiple access) that uses the basic DSSS (direct sequence spread spectrum) as its basis. However, there was also support for DSSS in accordance with the original 802.11 specification. While transmitting data, 802.11b uses the CSMA-CA technique that was defined in the original 802.11 base standard and retained for 802.11b.

The other standard defined almost at the same time was 802.11a and this standard used a 5-GHz ISM band and a different modulation technique, orthogonal frequency division multiplexing (OFDM) which enabled it to transfer raw data at a maximum rate of 54 Mbps. The data rate can be reduced to 48, 36, 24, 18, 12, 9 then 6 Mbps, if required, using different modulation techniques (such as BPSK) and coding rate. A more realistic data rate normally used with this standard is close to 20 Mbps. 802.11a has 12 non-overlapping channels, 8 dedicated to indoor and 4 for point-to-point communication.

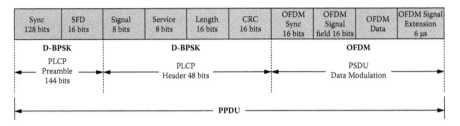

FIGURE 2.3
IEEE 802.11 PPDU frame

Among the above two standards, 802.11b became the main Wi-Fi standard and has been in use for some time.

However, with the advent of wireless systems, it was observed that one of the major shortfalls for developing applications operating on a Wi-Fi network is that it is not possible to allocate a required quality of service for a particular application. IEEE 802.11e was defined to address the quality of service or QoS problem. To introduce the QoS identifier, it has been necessary to develop a new MAC layer and this has been undertaken under the standard IEEE 802.11e. In this standard, the MAC layer has been modified to include the required functions for QoS management.

Another standard, 802.11g, was introduced in 2003. It uses a 2.4-GHz band and OFDM, and offers raw data rates of 54 Mbps. In addition, it also uses DSSS. The standard offers backward compatibility to 802.11b. 802.11g brings change in the structure of the data packet initially defined in the 802.11 standard. For Wi-Fi systems, the data packets sent over the radio interface consist of two parts: (i) a preamble that alerts receivers that a transmission is to start and then it enables them to synchronise; the preamble also contains information about the data to follow including the length of the payload; (ii) a payload which is the actual data sent across the radio network and can range from 64 bytes up to 1500 bytes. In most cases, the preamble and header are sent using the same modulation format as the payload, but this is not always the case. When using the DSSS-OFDM format, the header is sent using DSSS, while the payload uses OFDM.

Original 802.11 defines a long preamble frame set at the PLCP sublayer. In the 802.11b standard, an optional short preamble was defined. Then for 802.11g, the short preamble PPDU (PLCP protocol data unit) was defined as mandatory. Figure 2.3 shows a DSSS/CCK PPDU frame. The service field is always set to 0. The 802.11 standard reserves this space for future use.

Wi-Fi security is addressed in IEEE 802.11i which extends support for security schemes such as WEP, WPA and WPA2 with keys or codes provided for the various Wi-Fi hotspots in use.

TABLE 2.1
Summary of IEEE 802.11 standards

	802.11a	**802.11b**	**802.11g**	**802.11n**
Maximum data rate (Mbps)	54	11	54	600 (approx.)
Modulation	OFDM	CCK or DSSS	CCK, DSSS or OFDM	CCK, DSSS or OFDM
RF Band (GHz)	5	2.4	2.4	2.4 or 5
Number of spatial streams	1	1	1	1, 2, 3, or 4
Channel width (MHz)	20	20	20	20, or 40

IEEE 802.11n was proposed to provide better performance and to keep pace with the rapidly growing speeds provided by technologies such as Ethernet. The standard introduces the following major changes:

- Changes to implementation of OFDM - In addition to the *legacy mode* (a 20-MHz signal or a 40-MHz signal), two new modes are introduced. First, in the *mixed mode*, packets are transmitted with a preamble compatible with the legacy 802.11g and its remaining part has a MIMO training sequence format. Second, in the *Greenfield mode*, high throughput packets are transmitted without a legacy-compatible part.

- Introduction of MIMO, so that the multiple signal paths between a transmitter and receiver can be utilized to significantly improve the data throughput available on a given channel with its defined bandwidth.

- Power saving scheme for MIMO - Use of MIMO increases power consumption in the hardware circuitry. In order to reduce power consumption, when MIMO is not required (either when it remains idle or transmits at a very slow speed), the circuitry is held in inactive mode.

- Wider channel bandwidth - A 40-MHz channel bandwidth can optionally be used in addition to 20-MHz bandwidth in the previous systems. The choice is made dynamically by the devices.

- Antenna technology has been significantly improved with the introduction of beam forming and diversity.

- Reduced support for backward compatibility under special circumstances to improve data throughput.

802.11 Networks

The networks supported under this standard can either be infrastructured or may operate as ad hoc networks. Infrastructured networks are deployed inside buildings or to provide "hotspots" in areas like airport lounges. The entire area is split into a number of cells. Access points or base stations are installed which provide services to cells within the wireless network. A backbone wired network connects the access points to the servers.

An ad hoc network can also be set up when a number of computers (or other devices) are brought together and are required to be connected wirelessly for sharing data or accessing a peripheral. In this situation, there is requirement for establishing communication among the computers and not with a larger wired network. One of the devices takes over the role of master and controls the network, while other devices act as slaves.

2.3.2   IEEE 802.15.4

The IEEE 802.15.4 working group [14] defines a standard to specify the physical layer and media access control layer for wireless personal area networks (WPANs). The working group defined the standard in 2003. Later, the standard formed the basis for ZigBee, as well as 6LoWPAN which further extend the standard by developing the upper layers which are not defined in IEEE 802.15.4.

2.3.2.1   General Description

Usually, wireless personal area networks (WPANs) are infrastructureless and are used to transmit data only over short distances. IEEE 802.15.4 supports low rate WPANs (LR-WPANs) aiming at running industrial, residential and medical applications. The data rate and QoS requirements for such applications are not stringent and the low data rate enables the LR-WPAN to consume very little power. Thus, such networks work well with small, inexpensive, fixed or mobile devices requiring little or no battery power and typically operating within an operating space of 10 meters.

A device in an LR-WPAN network can either be a full-function device (FFD) or a reduced-function device (RFD). An FFD can function as a coordinator or just as an ordinary device. An FFD can communicate with RFDs or other FFDs, while an RFD can communicate only with an FFD. Two or more devices communicating within a small (10 meters) operating space and communicating on the same physical channel constitute a WPAN. A WPAN must include at least one FFD which operates as the PAN coordinator.

An LR-WPAN usually operates in either of the two topologies' star topology or peer-to-peer topology. These two topologies are shown in Figure 2.4. In the star topology, one of the FFDs is designated as the primary controller and is called the PAN coordinator. This device can be used to initiate, terminate, or route transmission around the network. All devices in the network

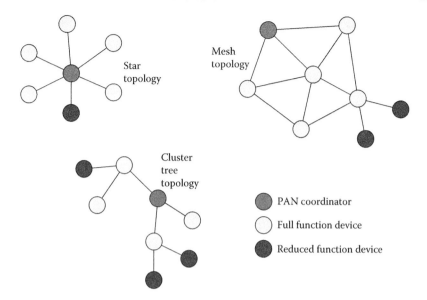

FIGURE 2.4 (SEE COLOR INSERT)
Three topologies of LR-WPAN

communicate only with the PAN coordinator. A PAN coordinator is generally mains powered, while other devices can be battery powered.

Though a device is designated as a PAN coordinator in peer-to-peer topology as well, unlike the star topology, these devices can also communicate with each other as long as they are within the communication range of one another. Thus, many complicated networks can be formed using a peer-to-peer topology.

A cluster-tree network is a special case of a peer-to-peer network. In a cluster-tree network, most devices are FFDs and there are few RFDs which become leaf nodes at the end of a branch. An FFD, which synchronizes with other devices and coordinates, itself becomes a coordinator. Only one of these coordinators is chosen as the PAN coordinator.

As mentioned earlier, IEEE 802.15.4 defines the specifications for two layers for LR-WPAN devices: the physical layer (PHY), which contains the radio frequency (RF) transceiver along with its low-level control mechanism, and a MAC sublayer that provides access to the physical channel for all types of transfer. A brief description of the two layers is given below.

2.3.2.2 Physical Layer

This layer offers activation and deactivation of the radio transceiver, energy detection (ED) within the current channel, link quality indication (LQI) for received packets, clear channel assessment (CCA) for CSMA-CA, channel frequency selection, data transmission and reception.

Sync Header		PHY Header		PHY Payload
Preamble	Start of Packet delimiter	Frame length		PHY service Data unit (PSDU)

FIGURE 2.5
Physical frame structure

A device can operate in three frequency bands: (i) the 2450-MHz band (2400–2483.5 range, O-QPSK encoding and 250 kb/s bit rate, used in Europe), (ii) the 915-MHz band (902–928 range, BPSK encoding and 40 kb/s bit rate, used in North America), and (iii) the 868-MHz band (868–868.6 range, BPSK encoding and 20 kb/s bit rate, used worldwide). A total of 27 channels, numbered 0 to 26, are available across these bands, where 16 channels are available in the 2450-MHz band, ten channels are available in the 915-MHz band, and one channel is available in the 868-MHz band. Thus, the standard provides two specifications, an 868/915-MHz direct sequence spread spectrum (DSSS) PHY which supports 20-kb/s and 40-kb/s over-the-air data rates and a 2450-MHz DSSS PHY, which supports 250-kb/s over-the-air data rate. A PHY can be chosen on the basis of the local regulations and user preferences. The raw data rate can be scaled down to fulfill the needs for sensor devices, thereby increasing the communication range of the devices.

The physical frame structure consists of a synchronization header (SHR) containing the preamble sequence and start-of-frame delimiter (SFD) fields, a PHY header (PHR) containing the length of the physical payload in octets and the PHY payload containing the *PHY service data unit* or PSDU. The preamble sequence enables the receiver to achieve symbol synchronization (Figure 2.5). The preamble sequence is 32 bits long, SFD field is 8 bits long, PHY header is also 8 bits and the length of the PSDU is less than or equal to 127 bytes.

2.3.2.3   MAC Layer

The MAC sublayer provides services for beacon management, channel access, frame validation, acknowledged frame delivery, association and disassociation of the devices within a PAN. Additionally, this layer also supports implementation of security mechanisms suitable for applications. The CSMA-CA mechanism is used for channel access. Primarily two services are provided at the MAC sublayer: MAC data service and MAC management service. The MAC data service enables transmission and reception of MAC protocol data units (MPDUs) across the PHY data service. MAC management service provides management commands between the next higher layer and the MAC layer.

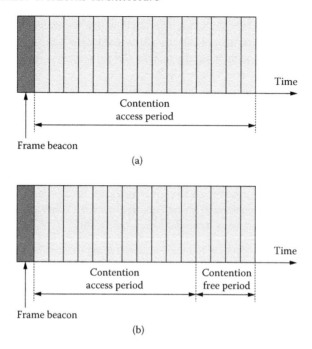

FIGURE 2.6
Superframe structure (a) without GTSs, (b) with GTSs

The MAC layer of LR-WPAN supports the non-beacon and beacon-enabled operating modes. These modes are operated using a superframe structure.

Superframe Structure

A superframe structure is optionally used in IEEE 802.15.4. The format of a superframe is defined by the coordinator. The superframe is divided into 16 slots. When beacon-enabled mode is supported, the beacon frame is transmitted in the first slot of each superframe. Here beacons are used for different purposes, such as for synchronization of the attached devices, for identification of the PAN, and to describe the structure of the superframes. The next part of the superframe is a *contention access period* (CAP). A device wishing to communicate during the contention access period (see Figure 2.6) competes with other devices using a slotted CSMA-CA mechanism. All transactions must be completed before arrival of the next network beacon.

A portion of a superframe may sometimes be dedicated to a specific application requiring fixed data bandwidth. These portions are termed guaranteed time slots (GTSs). A GTS may occupy more than one slot period and this period is treated as the contention-free period (CFP). A PAN coordinator may allocate up to seven such GTSs. CFP always appears at the end of the superframe immediately following the CAP, as shown in Figure 2.6. However, all contention-based transactions must be complete before the CFP begins. Further, a device transmitting during a GTS must ensure that its transaction completes before the end of the CFP.

When a coordinator does not use a superframe structure, it turns off the beacon transmissions.

Data Transfer Models

The standard supports three types of data transfer transactions: (i) data transfer from a device to coordinator, (ii) data transfer from coordinator to a device, and (iii) data transfer between two peer devices. In the case of star topology, only the first two types of transactions can occur. Whereas, in peer-to-peer topology, all three types of transactions may occur.

- In the first case, when a device transfers data to a coordinator, it first listens for the network beacon. After finding the beacon, the device transmits its data frame, using slotted CSMA-CA, to the coordinator. In response to successful reception of data frame, the coordinator acknowledges it sending an optional acknowledgment frame. If the network is not beacon-enabled, the coordinator simply transmits its data frame using unslotted CSMA-CA to the coordinator.

  For a non-beacon enabled network, the device directly transmits data frame to the coordinator using unslotted CSMA-CA. The coordinator, in turn transmits an acknowledgement frame. However, the acknowledgement frame is optional.

- When the coordinator transmits data to a device, it sends an indication to the device through beacon transmission. The device periodically listens to the network beacon and if it gets the indication that a message is pending, it transmits a MAC command requesting the data, using slotted CSMA-CA. The coordinator may send an optional acknowledgment frame. It then sends the pending data frame using slotted CSMA-CA. The device sends an acknowledgment frame to acknowledge the data by transmitting and when the acknowledgement frame is received by the coordinator, the message is removed from the list of pending messages in the beacon.

  For a network which is a not beacon-enabled, the coordinator waits for a request from the device. A device, when ready to receive data, contacts the coordinator by transmitting a MAC command requesting the data, using unslotted CSMA-CA. The rate of sending such requests is application-dependent. The coordinator acknowledges the data request by transmitting an acknowledgment frame and if there is any pending data, the coordinator transmits the data frame, using unslotted CSMA-CA, to the device. If there is no pending data, the coordinator transmits a data frame of zero-length payload. On successful reception of the data , the device transmits an acknowledgement frame.

- In peer-to-peer PAN, every device may communicate with every other device which remains within its transmission range. Thus, in such cases the devices wishing to communicate must constantly receive. Therefore, the device can simply transmit its data using unslotted CSMA-CA.

Frame Structures

The LR-WPAN defines the following four frame structures:

1. Beacon frames originate from the MAC sublayer. In a beacon-enabled network, coordinators transmit the beacon frames. The frame (i.e., MAC protocol data unit or MPDU) has three parts, a MAC header (MHR), a MAC service data unit (MSDU), and a MAC footer (MFR). The MAC service data unit (MSDU) contains the superframe specification, pending address specification, address list, and beacon payload fields. The MHR contains MAC frame control fields, beacon sequence number (BSN), and addressing information fields and finally MFR contains a 16-bit frame check sequence (FCS). It must be noted that two addressing modes are supported for the devices: 16-bit and 64-bit addressing modes.

   The MPDU is then passed to the PHY and becomes the PHY beacon packet payload (or PHY service data unit, PSDU). The PSDU is prefixed with SHR and PHR. The SHR, PHR, and PSDU together form the PHY beacon packet, (i.e., PPDU).

2. Data frames are generated at the upper layers. The data payload is passed to the MAC sublayer and is referred to as the MSDU. The MSDU is prefixed with an MHR and appended with an MFR. As in the previous case, the MHR contains the frame control, sequence number, and addressing information fields and the MFR contains the 16-bit FCS. The MPDU, which consists of MHR, MSDU, and MFR is passed to the PHY as the PHY data frame payload, (i.e., PSDU). The PSDU is prefixed with an SHR and a PHR forming the PHY data packet, (i.e., PPDU).

3. Acknowledgment frames originate from the MAC sublayer and consist of an MHR and an MFR. The MHR contains the MAC frame control and data sequence number fields. The MFR is composed of FCS. As in the earlier cases, the MPDU is passed to the PHY as the PHY acknowledgment frame payload, (i.e., PSDU) and is prefixed with the SHR and the PHR.

4. MAC command frames originate from the MAC sublayer. The MSDU of a MAC command frame contains the command type field and command-specific data. The MSDU is prefixed with an MHR and appended with an MFR.

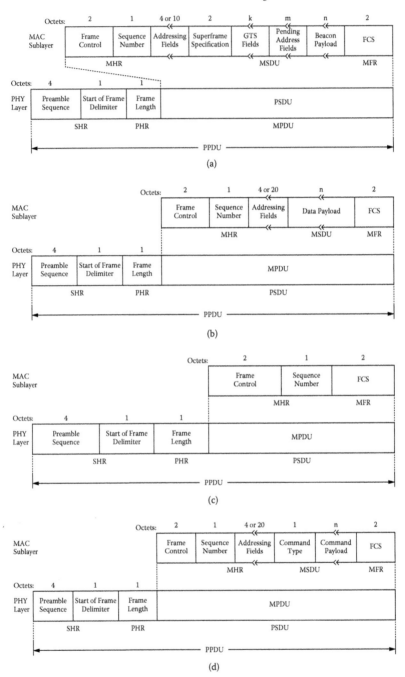

FIGURE 2.7
IEEE 802.15.4: (a) beacon frame, (b) data frame, (c) acknowledgment frame and (d) MAC command frame [14]

Addressing Modes

IEEE 802.15.4 allows IEEE 64-bit extended addresses or 16-bit addresses unique within the PAN after the device becomes associated with the PAN coordinator. Short addresses are assigned by the PAN coordinator during an association event; their validity and uniqueness are limited by the lifetime of their association. If either the association expires or the PAN coordinator stops functioning for any reason, the address is lost. The scalability and other issues posed by such a centralized allocation and single point of failure at the PAN coordinator also affects the growth of networks based on short addresses.

The LR-WPAN ensures robustness in the data transmission through implementation of certain mechanisms including the CSMA-CA mechanism, frame acknowledgment, and data verification using FCS. It is a lightweight protocol which is primarily applicable for battery-powered devices. These devices spend most of their operational life in a sleep state. Each device periodically listens to the RF channel to determine whether a message is pending for it. Thus, the devices enable duty cycling to reduce energy consumption. However, there are additional issues related to minimizing energy consumption and the application designer needs to handle these issues at other layers. We shall discuss these issues and techniques in subsequent chapters.

2.3.3 ZigBee

While IEEE 802.15.4 defines the physical layer and the MAC layer of LR-WPAN, ZigBee builds upon the IEEE 802.15.4 standard defines the network layer specifications and provides a framework for application programming in the application layer [2, 5].

2.3.3.1 Network Layer

Responsibilities of the ZigBee network layer include establishing a new network, joining or leaving a network, configuring new devices, assigning addresses to new devices joining a network, neighbor discovery and routing frames to their intended destinations.

The ZigBee network layer (NWK) supports the three topologies, star, tree, and mesh. In case of a star topology, a single device controls the network. The controlling device is called the ZigBee coordinator and all other devices are known as end devices. The coordinator is responsible for extending the network by assisting the end devices to join the network. The coordinator also manages the end devices on the network. The end devices directly communicate with the ZigBee coordinator. In mesh and tree topologies, the ZigBee coordinator is responsible for initiating the network and for deciding certain key network parameters. In certain situations, additional devices called routers are used for extending the network. Mesh networks allow peer-to-peer communication, whereas hierarchical routing strategies are implemented in tree networks.

ZigBee Devices

Three types of devices are supported in a ZigBee network layer: ZigBee coordinator, ZigBee router and end devices. ZigBee coordinators and routers are fully functional devices (FFDs as defined in IEEE standard). However, an end device can either be a full-function device (FFD) or a reduced-function device (RFD).

All devices are capable of joining and leaving the network anytime. ZigBee coordinators and ZigBee routers can assign logical network addresses to other devices and also maintain list of neighbouring devices. The coordinator is the only device capable of initiating a new network.

The coordinator also decides maximum number of routers, number of end devices that each router may have as children and maximum depth of the tree in case of tree topology.

Network Layer Services

Two types of services are provided at the network (NWK) layer. One is the NWK data service and the other is NWK management service. A general frame structure used by these services is shown in Figure 2.8 shows a general frame structure, the frame control fields, and structures of the command frame and the data frame.

The NWK management service provides primitives for a device (i) to discover the currently operating networks within its range, (ii) to join or rejoin a network, or to change the operating channel while remaining within an operating network, (iii) to synchronize or extract data from its ZigBee coordinator or router and (iv) to leave the network upon receiving a request from a higher layer. A device can start a new ZigBee network with itself as the coordinator, and permit other devices to join the network. In addition, a device designated as ZigBee router can also permit a device to join a network for a fixed period when it may accept devices onto the network. Primitives are also provided for a ZigBee router to initiate activities like routing of data frames and route discovery. Other primitives which are supported by ZigBee NWK management service include energy scan to evaluate channels in the local area and notifying the higher layers about network failures.

A NWK information base (NIB) is maintained by each device that comprises the attributes required to manage the NWK layer of a device. Each of these attributes can be read or written using NWK management service primitives. The read-only attributes cannot be written using primitives; rather they are set differently. For example, the time-out period for devices to retransmit a broadcast message or the maximum number of retries allowed after a broadcast transmission failure can be retrieved or set using these primitives. On the other hand, sequence number used to identify outgoing frames is a read-only attribute and is incremented every time the NWK layer sends a frame.

The NWK layer data service supports the transport of application protocol data units (APDUs) between peer application entities. Primitives are provided

Octets: 2	2	2	1	1	0/8	0/8	0/1	Variable	Variable
Frame control	Destination address	Source address	Radius	Sequence number	Destination on IEEE address	Source IEEE address	Multicast control	Source route subframe	Frame payload

NWK Header									Payload

(a)

Bits: 0–1	2–5	6–7	8	9	10	11	12	13–15
Frame type	Protocol version	Discover route	Multicast flag	Security	Source route	Destination on IEEE address	Source IEEE address	Reserved

(b)

Octets: 2	Variable	Variable
Frame control	Routing fields	Data payload
NWK header		NWK payload

(c)

Octets: 2	Variable	1	Variable
Frame control	Routing fields	NWK command identifier	NWK command payload
NWK header		NWK payload	

(d)

FIGURE 2.8
ZigBee frames: (a) general frame structure, (b) frame control, (c) data frame structure and (d) command frame structure

to transfer a data PDU (network service data unit) from the local APS sub-layer (application support sub-layer) entity (from the device itself) to a single or multiple peer APS sublayer entities (to other peer or child devices).

2.3.3.2 Application Layer

The ZigBee application layer is depicted in Figure 2.9. The application layer consists of an application support sublayer, ZigBee device object (ZDO) and application framework containing manufacturer-defined application objects.

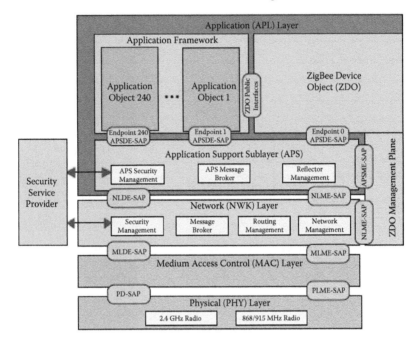

FIGURE 2.9
ZigBee application layer (Source: ZigBee Specification: ZigBee Alliance)

Application Support Sublayer

An interface between the network layer (NWK) and the application layer
(APL) is provided with the application support sublayer (APS). This layer extends support to the ZigBee application layer with APS data services and APS
management services. APS data services allow transmission of data between
two or more application entities residing within the same network. Fragmentation and reassembly of packets and provision of reliable data transmission
are also supported by these services. A variety of management services are
provided to application objects including security services and binding of devices. The sublayer also maintains a database of managed objects known as
the APS information base (AIB).

Application Framework

The application framework is an environment for hosting manufacturer-
defined application objects on ZigBee devices. Application objects (APOs)
encapsulate a set of attributes and allow the applications to set or retrieve
values of these attributes or be notified when an attribute value changes.
Within the application framework, 240 application objects can be defined,
each of which is identified by an endpoint address from 1 to 240. Two ad-

ditional endpoints are defined for the use of the data services: endpoint 0 is reserved for the data interface to the ZigBee device object (ZDO), and endpoint 255 is reserved for the data interface function to broadcast data to all application objects.

ZigBee devices staying within a network communicate among themselves with the help of agreements based on profiles. The types of profiles supported are *application profiles* and *device profiles*. Application profiles define messages, message formats and processes and create interoperable, distributed applications installed on separate devices. For example, devices deployed in a building can exchange messages to introduce cooperative control functions that enable automation, such as turning a lamp on or off and detecting movements of objects. Device profiles define common actions between ZigBee devices and are implemented by ZigBee device objects (ZDOs). Such actions include enabling autonomous devices to join a network and discover other devices and services on devices within the network.

ZigBee Device Objects

The ZigBee device objects (ZDOs) are responsible for initializing the application support sublayer (APS), the network layer (NWK) and the security service provider. These objects are also responsible for discovery, security management, network management, and binding management. The ZDOs are located between the application framework and the application support sublayer. These objects satisfy the common requirements of all applications operating in a ZigBee protocol stack.

ZigBee device objects (ZDOs) implement four primary communication functions between devices: device and service discovery, end device bind and unbind, binding table management and network management. While within a network, a ZigBee device can discover other ZigBee devices. In addition, using the service discovery process, it can also discover the capabilities of a given device within the same network. Service discovery can be accomplished by issuing a query for a given device or by using a match service feature (either broadcast or unicast).

Detailed specifications, interface descriptions, object descriptions, protocols and algorithms pertaining to the ZigBee protocol standard, including the application support sublayer (APS), the ZigBee device objects (ZDO), ZigBee device profile (ZDP), the application framework, the network layer (NWK), and ZigBee security services are further discussed in ZigBee Alliance specifications of NWK and APL layers [2].

2.3.4 6LoWPAN

6LoWPAN was introduced with a goal of enabling low-powered and low-data rate sensor devices communications using standard internetworking protocol.

It was first conceptualized by the Internet Engineering Task Force (IETF) [12] and implemented by several groups [6, 4, 11, 10].

In wireless personal area networks, sensor devices communicate using protocols like ZigBee implemented over IEEE Standard 802.15.4, Bluetooth, Wi-Fi etc. However, implementation of the most acceptable standard internetworking protocol, IP for wireless sensor networks, is considered necessary to ensure interoperability with all other potential IP network links, to increase robustness and to add flexibility to the system. Further, IP provides established security protocols for authentication, access control, and firewall mechanisms; allows functionalities for naming, addressing, translation, lookup and discovery; and supports proxy architectures for higher level services, such as load balancing, caching and mobility. It may also be noted that various application level data model and services have been developed over IP-based networks and most industrial standards support the IP option.

6LoWPAN offers an open interface implemented on top of IEEE 802.15.4 that allows use of Internet protocol (IP) as the common networking layer and transmission of IPv6 packets over low-power wireless personal area networks.

2.3.4.1   General Description

A LoWPAN (low-power wireless personal area network) communicates to other IP networks through one or more edge routers which forward IP packets or datagrams through wired or wireless media (Figure 2.10). In case of IP-based LoWPAN (as in 6LoWPAN), these edge routers forward datagrams at the network layer and therefore, there is no requirement for maintaining any application-layer state. Thus, unlike a ZigBee network, there is no requirement to use stateful and complex application gateways to connect LoWPANs to other networks. In ZigBee networks, application gateways communicate using application profiles and any changes to application protocols on the wireless nodes must also be accompanied with changes on the gateway. In contrast, IP-based edge routers remain unchanged in spite of changes in application protocols used in the LoWPAN [9].

Transmission of IPv6 packets on top of IEEE 802.15.4 is challenging. The major challenge is that the MTU size for IPv6 packets is 1280 bytes, which does not fit an IEEE 802.15.4 frame. The maximum physical layer packet size of IEEE 802.15.4 protocol data units is 127 bytes and maximum frame overhead is 25 bytes. Hence, the resulting maximum frame size at the medium access control layer becomes 102 bytes. Furthermore, since the IPv6 header is 40 bytes long, this leaves a smaller number of bytes for upper layer protocols and for application data. The problem becomes further complicated because IEEE 802.15.4 supports low-powered and low-throughput devices. So the network becomes prone to spurious interference, link failures, varying link qualities, and asymmetric links. In such situations, the network layer should take care of fragmentation and header compression and also data forwarding and routing. Moreover, IEEE 802.15.4 preferably supports a mesh topology

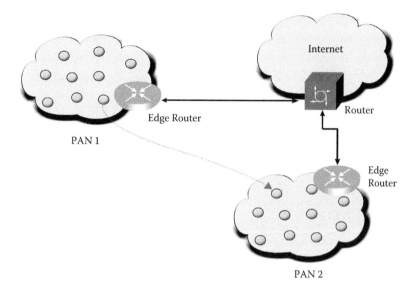

FIGURE 2.10 (SEE COLOR INSERT)
Personal area networks connected to Internet

with short-range connections. On the other hand, IP considers that the link is a single broadcast domain.

The above challenges have been handled in the 6LoWPAN recommendation by introducing an adaptation layer between the network layer and the data link layer, thereby allowing transportation of IPv6 packets over IEEE 802.15.4 links (Figure 2.11). The adaptation layer handles fragmentation and reassembling of IPv6 packets, as well as header compression [13].

The three main aspects introduced by the adaptation layer for implementing IP-enabled low-powered networks are described below.

- **Header compression:** As mentioned earlier, the IPv6 header is 40 bytes long. In addition to these 40 bytes, more space is used by upper-layer protocols, introducing more overhead in the datagram. For example, UDP uses 8 bytes in the header. However, it is evident that many of these header fields contain redundant information. Some common values are present in the headers at different layers. These header fields are omitted from a packet when they can be derived from lower-level layers or on the basis of simple assumptions of shared context.

- **Fragmentation:** One major issue is that IPv6 MTU requirement is 10 times the IEEE 802.15.4 frame size. Thus, IPv6 packets need to be fragmented into multiple link-level frames to accommodate the minimum MTU requirement.

- **Link-layer mesh routing under IP topology:** IEEE 802.15.4 subnets may utilize multiple radio hops per IP hop. Link-layer mesh routing or layer

| Application Layer |
| Transport Layer |
| Network Layer(IPv6) |
| 6LoWPAN Adaptation Layer |
| 802.15.4 MAC |
| 802.15.4 PHY |

FIGURE 2.11
Layered stack of 6LoWPAN

2 forwarding enables delivery of IPv6 datagrams over multiple radio hops in IEEE 802.15.4 PANs. To support mesh routing of IPv6 packets, the adaptation layer allows to carry link-level addresses for the ends of an IP hop.

Each of the above capabilities is expressed in a self-contained subheader. A header type field is placed at the beginning of each header to indicate the type of header. In the remaining part of this section, we discuss implementation of each capability along with the header formats.

2.3.4.2   Frame Format

6LoWPAN uses header stacking to handle each capability separately using a well-defined method. Thus, all LoWPAN datagrams transmitted using IEEE 802.15.4 are prefixed by an encapsulation header stack. Each header in the stack contains a type and fields expressing the capabilities.

The 6LoWPAN dispatch header, which is the first byte in IETF format, helps the receiver understand what is coming next. The first two bits indicate whether it is a LoWPAN frame or not and in case of a LoWPAN frame, it indicates whether the remaining part is a mesh header, or fragmentation header or IPv6 addressing header. The last two bits of this octate show whether the following bytes contain a HC1 compressed header or uncompressed IPv6 address header.

Header Compression Scheme

RFC 4944 [13] introduces a scheme HC1 that compresses the entire 40-byte IPv6 address header into only few bits optimized for link-local IPv6 communication. The fully compressed byte in HC1 follows the compressed IPv6 dispatch header. The compression scheme removes some fields from the upper-layer headers. For example, the network prefixes (64 bits) of *source and destination addresses* are reduced to a single bit each when they carry the link-local

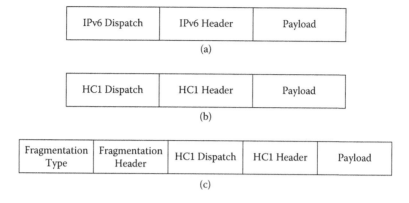

| IPv6 Dispatch | IPv6 Header | Payload |

(a)

| HC1 Dispatch | HC1 Header | Payload |

(b)

| Fragmentation Type | Fragmentation Header | HC1 Dispatch | HC1 Header | Payload |

(c)

FIGURE 2.12

6LoWPAN header stacking: (a) LoWPAN encapsulated IPv6 datagram, (b) LoWPAN encapsulated HC1 compressed datagram, (c) LoWPAN encapsulated compressed datagram requiring fragmentation

prefix with the assumption that they can be derived from link-addresses. The 64-bit interface identifiers (IIDs) for both *source and destination addresses* are deleted, because they can also be derived from the corresponding link-layer addresses in IEEE 802.15.4. The *next header* field is reduced to two bits whenever only one of the three higher level protocols, UDP, TCP, or ICMPv6, is used. *Traffic class and flow label* are compressed into a single bit when both have zero values. *Payload length* can be derived from IEEE 802.15.4 frame. *Version* is always deleted as only IPv6 is used. The *hop limit* field remains uncompressed.

Figure 2.13 shows an IPv6 header and its compressed form using the HC1 scheme. The first byte is the dispatch byte which indicates use of HC1. The next 8 bits describe how the IPv6 fields are compressed. One bit is used for both source and destination addresses to indicate that the IPv6 prefix is link-local and can be removed. One bit is used to indicate whether the interface identifier or IID can be derived from the IEEE 802.15.4 link address. The TF bit indicates whether traffic class and flow label are both zeros and can be eliminated. The two *next header* bits indicate the next header value as discussed earlier. The HC2 bit indicates whether the next header is compressed using the HC2 scheme. Thus, when it is fully compressed, the HC1 encoding reduces the IPv6 header to three bytes, including the dispatch header. *Hops left* is the only field which remains uncompressed.

For *UDP header compression* in RFC 4944, the HC2 bit is set in the HC1 encoding and an additional byte is included following the HC1 encoding bits to specify how the UDP header is compressed. A range of ports is used to effectively compress UDP ports. When the ports fall in this range, the upper bits are removed. If both ports, source and destination, fall within range, their

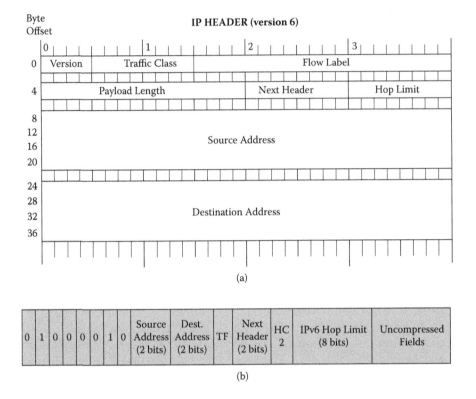

FIGURE 2.13
Header compression: (a) IPv6 header, (b) IPv6 header compression

addresses can be compressed down to a single byte. HC2 also allows deletion of the UDP length, as it can be derived from the IPv6 payload length field.

Handling Fragmentation

The IPv6 payload is fragmented, so that each fragment can fit into a single IEEE 802.15.4 frame. For handling the fragmented IPv6 payload, a *fragment header* is used by the 6LoWPAN standard. This fragment header includes three fields, (i) datagram size, (ii) datagram tag, and (iii) datagram offset. Fragments are assigned sequentially at the source of fragmentation. However, they do not have to arrive at the receiver in order. Datagram size provides the total size of the unfragmented payload and is included in every fragment, so that the receiver can allocate buffer when fragments arrive out of order. The datagram tag identifies the set of fragments that correspond to a given payload and is used to match fragments of the same payload. Thus, all fragments of the same IP packet should bear the same tag. Datagram offset identifies the

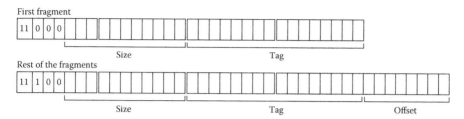

FIGURE 2.14
6LoWPAN fragment header

position of the fragment within the unfragmented payload in 8-byte chunks. Figure 2.14 shows the 6LoWPAN fragment header format.

Mesh Addressing

6LoWPAN allows link-layer mesh routing on a IP-enabled network. Thus, IEEE 802.15.4 subnets may utilize multiple radio hops per IP hop. A *mesh addressing* header is used to forward 6LoWPAN payloads over multiple radio hops. The header includes three fields, (i) hop limit, (ii) source address, and (iii) destination address. The hop limit field defines the number of hops for forwarding. It is decremented by each forwarding node, and the frame is dropped when the value in this field becomes zero. The source and destination addresses indicate the end points of an IP hop. Both are IEEE 802.15.4 link addresses and may carry either a 16-bit short or IEEE 64-bit extended address.

The format of the mesh addressing header is shown in Figure 2.15. The header type is only two bits which indicate that the remaining part contains the mesh header. Next two bits indicate which addressing mode is used for the source and destination addresses. The following four bits carry the hop limit and the remaining part contains addressing fields. The mesh addressing header can have length between 5 and 17 bytes, depending on the addressing modes in use.

FIGURE 2.15
6LoWPAN mesh addressing header

Addressing Modes

IPv6 nodes are assigned 128 bit IP addresses in a hierarchical manner through an arbitrary length network prefix. IEEE 802.15.4 devices may use either IEEE 64-bit extended addresses or, after an association event, 16-bit addresses that are unique within a PAN. There is also a PAN-ID for a group of physically collocated IEEE 802.15.4 devices.

RFC 4944 [13] recommends that an IPv6 interface identifier [7] for an IEEE 802.15.4 interface may be based on the 64-bit address assigned to the IEEE 802.15.4 device when the device has an IEEE EUI-64 address. But when a device has a 16-bit short address, a "pseudo 48-bit address" is formed first, by concatenating 16 zero bits to the 16-bit PAN ID, and then these 32 bits are concatenated with the 16-bit short address. The interface identifier is formed from this 48-bit address as per the IPv6 over Ethernet specification. The IPv6 link-local address for an IEEE 802.15.4 interface is formed by appending the interface identifier, as defined above, to the prefix FE80::/64.

2.4 Summary

The chapter summarizes different communication standards used in a wireless sensor network. The default standard used by most sensor devices is IEEE 802.15.4. The IEEE 802.15.4 working group specifies the physical layer and media access control layer for wireless personal area networks. Later, a proprietary protocol ZigBee was built upon IEEE 802.15.4. ZigBee provides specifications for network and application layers. 6LoWPAN offers an open interface to allow use of IP as the common networking layer and transmission of IPv6 packets over low-power wireless personal area networks. An adaptation layer is introduced between 802.15.4 MAC layer and IPv6 network layer for transmission of IPv6 packets. In addition, IEEE 802.11 standards are also discussed as sensor devices are often integrated with wireless LANs based on IEEE 802.11.

Bibliography

[1] I.F. Akyildiz, W. Su, Y. Sankarasubramaniam, and E. Cayirci. Wireless sensor networks: a survey. *Computer Networks*, 38(4):393–422, 2002.

[2] ZigBee Alliance. ZigBee specification. Document 053474r17, 2008.

[3] IEEE Standards Association. Website: http://standards.ieee.org/about/get/802/802.11.html.

[4] Berkeley WEBS: Blip. Website: `http://smote.cs.berkeley.edu:8000/tracenv/wiki/blip`.

[5] S. C. Ergen. ZigBee/IEEE 802.15.4 summary. Website: `http://www.eecs.berkeley.edu/csinem/academic/publications/zigbee.pdf`, 2004.

[6] M. Harvan. Connecting wireless sensor networks to the Internet—a 6Lowpan implementation for tinyos 2.0. *Jacobs University Bremen, Germany*, 2007.

[7] R. M. Hinden and S. E. Deering. Internet protocol version 6 (ipv6) addressing architecture (rfc 4291). February 2006.

[8] M. Holland, T. Wang, B. Tavli, A. Seyedi, and W. Heinzelman. Optimizing physical-layer parameters for wireless sensor networks. *ACM Transactions on Sensor Networks (TOSN)*, 7(4):28, 2011.

[9] J. Hui, D. Culler, and S. Chakrabarti. GLowPan: Incorporating IEEE 802.15.4 into the IP architecture.

[10] Contiki 2.6 Implementation. Website: `http://contiki.sourceforge.net/docs/2.6/a01794.html#_details`.

[11] Blip: Berkeley IP implementation for low-power networks. Website: `http://tinyos.stanford.edu/tinyos-wiki/index.php/BLIP_2.0_Tutorial`.

[12] N. Kushalnagar, G. Montenegro, and C. Schumacher. IPv6 over low-power wireless personal area networks (6LowPANs): Overview, assumptions, problem statement, and goal (RFC 4919). 2007.

[13] G. Montenegro, N. Kushalnagar, J. Hui, and D. Culler. Transmission of ipv6 packets over IEEE 802.15.4 networks (rfc 4944). September 2007.

[14] LAN-MAN Standards Committee of the IEEE Computer Society. 802.15.4: Wireless medium access control (MAC) and physical layer (PHY) specifications for low-rate wireless personal area networks (LR-WPANs), 2003.

[15] G. J. Pottie and W. J. Kaiser. Wireless integrated network sensors. *Communications of the ACM*, 43(5):51–58, 2000.

[16] J. H. Schiller. *Mobile Communications*. Pearson Education, 2003.

[17] E. Shih, S. H. Cho, N. Ickes, R. Min, A. Sinha, A. Wang, and A. Chandrakasan. Physical layer driven protocol and algorithm design for energy-efficient wireless sensor networks. In *Proceedings of the 7th annual international conference on Mobile computing and networking*, pages 272–287. ACM, 2001.

[18] K. Sohrabi, B. Manriquez, and G. J. Pottie. Near ground wideband chan-
     nel measurement in 800-1000 MHZ. In *Vehicular Technology Conference,
     1999 IEEE 49th*, volume 1, pages 571–574. IEEE, 1999.

[19] K. D. Wong. Physical layer considerations for wireless sensor networks. In
     *Networking, Sensing and Control, 2004 IEEE International Conference
     on*, volume 2, pages 1201–1206. IEEE, 2004.

3

Information Gathering

3.1 Introduction

The importance of a wireless sensor network lies in its ability to sense an event and propagate the sensed data to the sink through a multi-hop network. In a typical wireless network like WiFi or a cellular network, nodes directly communicate with the nearest high-power control tower or base station. However, in a wireless sensor network, sensor nodes communicate only with their local peers. Thus, instead of relying on a pre-deployed infrastructure, each individual sensor node becomes part of the overall infrastructure. This enables the sensor network to spread over a large geographical area. In a cellular network, services may be denied when too many cell phones are active in a small area. In contrast, if nodes are added to a wireless sensor network, interconnection between the nodes only grows stronger. As long as the node density is sufficiently high, a single network can grow to cover an unlimited area.

The structure of wireless sensor network as described above can function efficiently with the implementation of peer-to-peer networking protocols which provide a mesh-like interconnect and support transmission of data between the thousands of sensor nodes in a multi-hop fashion. A large number of multi-hop routing algorithms have been designed for effective transmission of sensed data in a wireless sensor network. Some important routing algorithms are discussed in Section 3.2.

Many routing algorithms have been designed to rely on the location information of the nodes and on the basis of this information these algorithms make packet forwarding decisions. Such algorithms are categorized as *geographical routing* and are discussed the Section 3.3.2. In order to implement geographical or location-based routing, the sensor nodes must possess information about their locations. Techniques and schemes for making the nodes location-aware are discussed in Section 3.3.1.

When applications sense specific data and forward the data to the sink through a multi-hop network, simple routing or data forwarding algorithms can be implemented. But in some applications like environment monitoring, large numbers of sensor nodes generate data and for precise results all these data must be analyzed. However, propagation of such a large volume of data through the network not only creates congestion in the network, it also increases energy consumption, thereby reducing network lifetime. Thus, there is a need for reducing data flow to increase the efficiency of the applications and the network lifetime. Fortunately, in a dense network, various sensor nodes often detect common phenomena. Therefore, there are redundancies in the data that various nodes communicate to the sink. Such redundancies must be removed during forwarding of data toward the sink. Furthermore, in-network filtering and processing techniques can be implemented to reduce the data flow towards the sink. The techniques also help to conserve the energy resources. Data aggregation techniques are embedded in the routing protocols in wireless sensor networks to achieve the above goal. Section 3.4 gives an overview of data aggregation.

Routing algorithms and data aggregation algorithms in the context of energy efficiency are discussed in Chapter 4.

Data aggregation techniques are implemented to combine the data coming from different sources, eliminate redundancies and minimize number of transmissions towards the sink. In order to achieve the basic objectives of in-network processing, the traditional focus on address-centric approaches (transmitting data from an addressable node to another addressable node through a shortest route) in a network has shifted to a data-centric approach (finding routes from multiple sources to a single destination that allows in-network consolidation of redundant data). Naming schemes in wireless sensor networks are necessary for appropriate implementation of data-centric approaches. The requirement for applying naming schemes is briefly discussed in Section 3.5.

3.2   Routing

Many state-of-the-art routing algorithms for wireless sensor networks have been designed following the traditional topology-based approach, i.e., forwarding decisions are based on the neighbor node information or in other words on the currently available links between network nodes.

Some of the algorithms proposed in the early days of wirless networks are proactive, i.e., every node maintains tables that store the entire topology of the network. The tables are updated regularly to maintain valid routing information from each node to every other node at any instance of time. These algorithms are not scalable because of the increasing table size for a large network. Moreover, topology information needs to be exchanged between the nodes on a regular basis and such exchanges are responsible for high overhead on the network. Therefore, proactive routing protocols are not suitable in a sensor network.

In contrast, reactive protocols seek to set up routes on demand. When a node requires initiation of communication with another node, the routing protocol tries to establish such a route. However, in case of reactive protocols, a large amount of overhead may be incurred due to route discovery and setting up routes in a dynamic manner. Many algorithms use a combination of both.

Wireless sensor network routing algorithms are broadly classified as *flat-based* and *hierarchical* [1].

3.2.1 Flat-based Routing Algorithms

Flat routing algorithms assume that all nodes in the sensor network play the same role. These nodes collaborate to forward the data from the source nodes to some other intended nodes (including the sink) through multiple hops. Usually, these algorithms are data-centric and are implemented as either push-based or pull-based or a combination of the two.

The simplest approach to data dissemination in the network is *flooding*. *Classical flooding* starts from a source node. The source node that senses the data broadcasts it to all its neighbors. When a node receives the data, it stores a copy and, in turn, forwards it again to its neighbors. This is a straightforward and fast technique for data dissemination and forwarding the data to the sink. However, the approach suffers from three deficiencies [15]:

- Implosion - This problem occurs when a node receives two copies of the same data from two different neighbors through different routes. The system wastes energy and bandwidth for this redundant movement of data.

- Overlap - Two sensors may cover overlapping geographical regions. When the sensors forward their data to a node, the node receives two copies of the same data.

- Resource blindness - In case of classical flooding, sensor nodes do not modify their activities based on the amount of resources they have. For example, a node with very low energy can repeatedly be used as a data forwarding node in spite of the risk of taking this node close to death due to energy scarcity. This is one of the major drawback in an energy-constrained network.

Gossiping is an alternative to the classical flooding approach that uses randomization at the time of data forwarding. At every step, each node only forwards data to one neighbor selected randomly. The major problem is that data may never reach some nodes.

As discussed earlier, wireless sensor networks have no router and only hop-by-hop communication is carried out. Therefore, no globally unique identifier is needed and only the neighbours must be distinguishable. In order to overcome the problems of *flooding* and *gossiping*, push-based and pull-based routing protocols have been proposed.

3.2.1.1 Sensor Protocols for Information Negotiation (SPIN)

SPIN [15] is one of the popular routing protocols in the wireless sensor networks. It is a push-based protocol. When a node senses data (or possesses a piece of data from other nodes), it advertises the data to its neighbors. If any neighboring node is interested, a request is sent to the source and the source (or the owner of the data) in turn sends the data.

SPIN is basically a negotiation-based protocol which treats all sensors as potential sink nodes. Such negotiation ensures that only useful information will be transferred. The negotiation takes place with the use of metadata and exchange of messages.

Metadata

Each piece of data is named or described using metadata. Thus, same sensor data has the same metadata and two different sensor data have different metadata. Thus, a metadata x distinctly identifies a piece of data X and distinguishes it from other data. The size of x must be much shorter than X. Some overheads are associated with the storage, retrieval and general management of metadata. However, because the size of this metadata is smaller than the actual data, and actual data is sent only when it is required, the communication cost is reduced by a large amount.

SPIN does not specify a format for metadata. The format is usually application-specific and is defined within the application. For example, (x,y,ϕ) may be a metadata for a camera sensor at geographic coordinates (x,y) with orientation ϕ. Applications are also responsible for interpreting or synthesizing the metadata.

Messages

SPIN uses three types of messages: (i) ADV (for advertising new data), (ii) REQ (for sending requests for data), and (iii) DATA (data messages containing actual sensor data).

When a node has data to share, it advertises the fact by transmitting an ADV message. The ADV message contains only metadata. Thus, its size is small. Once a node receives an ADV message, it checks whether it has any interest in that data. If it is interested, then the node sends an REQ message to the node that advertised the data. The REQ message contains only metadata. Finally, the node that possesses the data sends a DATA message containing actual sensor data with a metadata header to the requesting node. ADV and REQ messages contain only metadata and therefore are smaller and cheaper to transmit.

Versions of SPIN

Four different versions of SPIN protocol have been proposed [15].

1. SPIN-PP is designed to perform optimal point-to-point communication. Two nodes can take part in exclusive communication with each other without interference from the other nodes. Thus, the cost of communication with *n* nodes in terms of time and energy

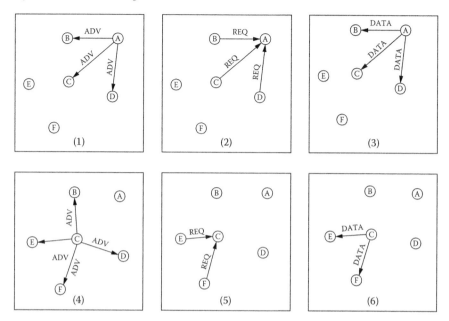

FIGURE 3.1
SPIN-PP routing in WSN

is n times more expensive than communicating with one node. The same three-way handshaking protocol is used for SPIN-PP.

Because of point-to-point communication, if node B receives an ADV message from node A, it does not have to send an ADV message to A along with its neighbors. Node B can aggregate its own data (if it possesses data) with the data from node A and send advertisement for the aggregated data to its neighbors. This protocol is suitable for lossless networks.

Major advantage of SPIN-PP is its simplicity. Each node requires information only about its single-hop neighbors and does not require other topology information. Figure 3.1 depicts the protocol.

2. SPIN-BC is designed for broadcast networks in which the nodes use a single shared channel to communicate. Thus, when a node sends a message, it is received by all other nodes within a certain range of the sender.

   The source node, before transmitting the messages, first senses the channel to check whether it is free. Next the ADV message is broadcast. A node which receives an ADV message first checks whether it has already received or requested for the advertised data. If not, a randomly chosen timer (for a predetermined interval) is set. When the timer expires, the node broadcasts a REQ message containing

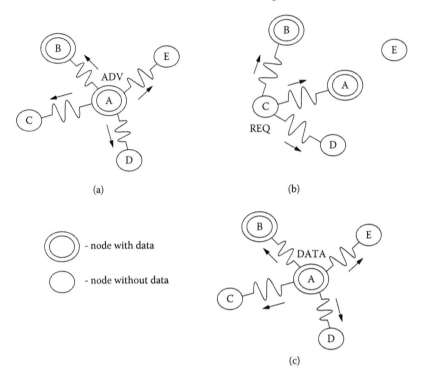

FIGURE 3.2
SPIN-BC routing in WSN

the id of the original advertising node mentioned in the message header. When a node other than the advertising node receives the REQ message before expiration of the timer, it cancels its own request (if there is any). Thus, possibilities of redundant requests for the same message are reduced. When the advertising node receives an REQ message, it broadcasts the DATA message only once and it is received by all the requesting nodes within the range. Even though a node may receive multiple requests for the same message, it sends the DATA message only once.

SPIN-BC is based on one-to-many communication which is cheaper than point-to-point communication. Further, multiple requests are handled with a single broadcast and redundant requests are suppressed. This saves communication time and energy. Figure 3.2 explains the protocol in detail.

3. SPIN-EC is the same as SPIN-PP except than an energy consideration heuristic is included. Thus, a node participates in the negotiation only if it contains energy above a certain level defined by a

threshold. The node should be able to complete all the stages of the protocol. In response to an ADV message, a node sends an REQ message only if it has enough energy to transmit an REQ message and receive the corresponding DATA message.

4. SPIN-RL is a reliable version of SPIN-BC for lossy networks. Here, each node keeps track of all the advertisements it receives and the corresponding source nodes. If it does not receive a piece of data it has requested within a certain time, it sends the request again. Nodes also put a limit on the frequency with which the data messages can be resent. After sending the DATA message, a node waits for a certain period before it responds to other requests for the same DATA message.

SPIN protocol overcomes the implosion problem as transmission of redundant data messages can be eliminated through negotiation. When a node has particular data, it never sends request for it. SPIN also overcomes the problem of overlapping and adapts to the current resource availability scenario.

3.2.1.2 Directed Diffusion

Directed Diffusion [12] is a pull-based approach. A node (destination or sink) which is interested in receiving some data initiates a query message with the goal of establishing a route between the source and the destination. Following this *interest message*, the sink node and other nodes (including a source node) participate in transmitting messages until a preferable route is established from source to sink and data messages are forwarded via this route.

Four types of messages are generated and transmitted during routing using Directed Diffusion technique: (i) *interest message*, (ii) *exploratory message*, (iii) *reinforcement message*, and (iv) *data message*. While, interest messages and exploratory messages are broadcast, other two types of messages are sent as unicast messages.

First, the destination node broadcasts its interest message for a piece of data. The interest message, when received by a neighbor of the querying node, is again broadcast to all the neighbors of the node. This way demand for a sensor data is propagated through the network. Interest of a node is expressed as a list of attribute-value pairs which describe an information gathering task. Examples of query messages are "track the animal x" or "get the temperature of the north-east corner of the room" etc.

Each node remembers which neighbour(s) sent the interest and sets up a gradient towards that neighbour. Gradient represents the direction towards which data matching an interest flows and also status of the demand generated at the sink (represented as the interest message).

Every node stores and interprets interests. An interest cache is maintained at every node. An item is stored in the cache corresponding to an interest expressed by the sink node that has reached the node. When a sensor node finds that it can match the interest, it activates the local sensors to collect data and becomes the source of data.

The source node generates data message matching the interest and broadcasts the data message to all neighbors for which it has matching gradients. Data messages are generated using the same attribute-based naming scheme which is used for interest messages. The initial data message from the source is marked as an exploratory message. The message is forwarded through the network in the same way as the interest message and finally reaches the sink. If the sink receives the exploratory message from multiple neighbors, it chooses to receive subsequent data messages for the same interest from a preferred neighbor (for example, the one that delivered the first copy of the data message).

The sink reinforces the preferred neighbour by sending a unicast reinforcement message. The neighbor node in turn reinforces its preferred upstream neighbour and this goes on until the source node is reached. Thus, a route is eatablished from source to sink.

After the initial exploratory data message, subsequent messages from source are sent only on the reinforced path. While propagating towards the sink, data is stored in the interest cache of every intermediate node. Cached data can be used to discard duplicate data and to prevent loops.

Figure 3.3 shows the four steps of the Directed Diffusion routing algorithm. Using this routing scheme, on-demand route setup is possible. The scheme can be improved to incorporate in-network processing. Thus, each node can perform aggregation along with caching.

Directed Diffusion is a query-driven scheme. It incurs additional overhead for data matching and generating queries. Thus, it is not a good choice for applications requiring continuous data delivery.

3.2.1.3   Rumor Routing

Directed Diffusion sends a query requesting a piece of data and the query is flooded within the network in the search of a shortest path between the source of data and the sink. On the other hand, SPIN propagates the event information in the network through advertisement using metadata. In contrast with the above two algorithms, Rumor Routing [4] proposes flooding the event information in the network. It is basically a logical compromise between flooding queries and flooding event notifications. Rumor Routing does not guarantee the shortest path between the source and the destination. Thus, it is suitable for applications that have no such requirements. For example, if applications require transmission of data of short length, flooding every query may incur large overhead in comparison with the cost of delivering the data via a non-optimal route. Rumor Routing may be implemented in such cases.

In the case of Rumor Routing, each node maintains a list of its neighbors and an events table. A node may actively broadcast a request to get the ids

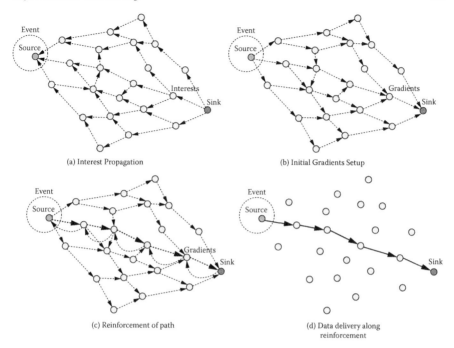

(a) Interest Propagation

(b) Initial Gradients Setup

(c) Reinforcement of path

(d) Data delivery along reinforcement

FIGURE 3.3
Directed Diffusion routing in WSN

of its neighbors. A node can also passively listen to the broadcast messages of other nodes in order to create and maintain a neighbor list.

The event table stores information about all the events known to the node and gradients towards the events including distances in the number of hops. When a node witnesses an event, it adds the event to its event table, with a distance zero to the event. It also probabilistically generates an agent. An agent is a long-lived packet generated for flooding the local event information in the network. Along with the local event information, the agent carries an event table that stores a list of all the events that the agent has encountered on its route. Thus, when an agent arrives at a node, it synchronizes its own table with the event table stored at the node. All events known to the agent are added to the event table stored at the node. The agent also updates its own table with the events listed in the event table stored at the node. In case an event is known to both, the agent and the node, the paths to the event are updated and after table synchronization is complete, both event tables contain the best route to that event. After table synchronization, the time-to-live (TTL) value of an agent is decremented, and if it is greater than zero, the agent is forwarded. An agent travels the network for some number of hops (based on its TTL value) and then dies.

All transmissions in Rumor Routing are broadcasts and therefore the neighboring nodes within the range of transmission can hear the agent. These nodes in turn can modify their event table to include all the events that the agent knows about. An agent also maintains a list of recently seen nodes. Thus, it can eliminate the chance of revisiting a node and formation of a loop.

Any node may also generate a query for a particular event. If the node already has information about the route to the event, the query is transmitted directly to the event. Otherwise, the query is forwarded in a random direction. This continues until TTL value of the query packet expires or until the query reaches a node that has either observed or received information about the target event. If a loop is detected by a node, the query packet is not forwarded further and it dies.

If the node that originated the query finds that the query has not reached the event source for some reason (the query packet was lost or its TTL value became zero), it can try retransmitting or flood the query.

Node A in Figure 3.4 knows a route to event $E1$ and event $E2$. Agent 2 knows a route to event $E1$ only. When the agent visits the node A, it updates its event table and includes information about $E2$. It also updates the event table at the node, because it knows a shorter route to event $E1$ via node B.

3.2.2   Hierarchical Routing Algorithms

Flat-based routing is useful when the nodes are homogeneous in terms of resources. However, when some nodes are more powerful than others in terms of computational capabilities, storage, energy, etc., a hierarchical architecture of the network can be established in which powerful nodes are used to process and send the information while less powerful nodes are used to perform the sensing tasks. A hierarchical structure can be established by creating clusters and assigning special tasks to the cluster-heads. Two major phases are defined for hierarchical routing techniques—first selection of clusterheads and then routing information in the network.

This section covers some of the popular hierarchical routing algorithms.

3.2.2.1   LEACH Routing Protocol

LEACH (low-energy adaptive clustering hierarchy) [11, 10] routing is a hierarchical protocol that considers that the base station is fixed and located far from the sensor nodes and all nodes in the network are homogeneous and energy-constrained. Thus, transmission of data directly from sensor nodes to the base station is expensive. Therefore, only a few nodes known as cluster-heads forward the sensed data to the base station. All other nodes form clusters with respective cluster-heads, sense the data in their proximity and transmit the sensed data to the cluster-heads. The cluster-heads are also capable of processing data using *data aggregation* or *data fusion* algorithms, so that less

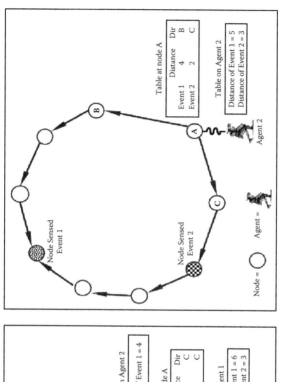

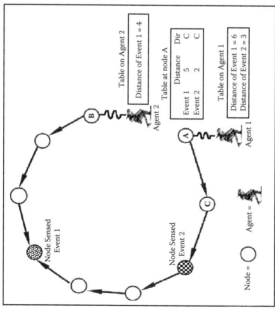

FIGURE 3.4
Rumor routing in WSN

data is required to be transmitted to the base station. Unlike other clustering algorithms, LEACH allows adaptive clusters and rotating cluster-heads.

The LEACH algorithm operates in different phases. First, there is a set-up phase when the clusters are organized followed by a steady-state phase, when sensed data is transferred to the base station. Steady-state phase is longer than the set-up phase. The two phases are described below:

Advertisement Phase

During this phase, selection of the cluster-heads is initiated. Each node can become a cluster-head. This decision is made by a node only if it has not become a cluster-head in the last $1/p$ rounds. The decision is based on the desired percentage of cluster-heads (decided a priori) and the current round. A threshold value is computed based on the above parameters. Each node chooses a random number between 0 and 1. If the number is less than the threshold value, the node decides to be a cluster-head.

During the first round (round 0), each node can become a cluster-head with probability P (desired percentage of cluster-heads). The nodes that are cluster-heads in round 0 cannot be cluster-heads for the next $1/p$ rounds. Therefore, probability of the remaining nodes for becoming cluster-heads is increased, since fewer nodes are eligible to become cluster-heads. After P rounds, all nodes again become eligible to become cluster-heads with probability P.

A node that decides to become a cluster-head for the current round broadcasts an advertisement to the rest of the nodes. Non-cluster-head nodes within the range of transmission of any of the cluster-head nodes listen to the advertisements. A node may receive advertisements from more than one cluster-head.

Set-up Phase

After the above phase is complete, each non-cluster-head node decides the cluster it wants to join. This decision is based on the received signal strength of the advertisement. Once the cluster is chosen by each node, the node informs the corresponding cluster-head node. All cluster-heads receive this information and form the cluster. The cluster-head node also creates a TDMA schedule based on the number of nodes in the cluster. The schedule is broadcast to the nodes in the cluster that informs each node when it can transmit.

Data Transmission

After the set-up phase, data transmission begins in the network. Each node can send data to the cluster-head only during its allocated transmission time following the schedule. In the remaining period, the radio of each non-cluster-head node is turned off. The cluster-head node keeps its receiver on to receive all the data from the nodes in the cluster. When all the data are received,

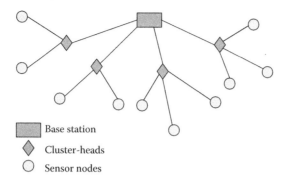

FIGURE 3.5
Hierarchical structure in a network

the cluster-head node may apply some algorithm for processing the data. The processed data is finally sent to the base station. The LEACH algorithm assumes that the base station is far away. Therefore, cluster-head to base station transmission is a high-energy transmission.

Data transmission phase is the steady-state phase for the LEACH protocol. After completion of this phase (duration is determined a priori), the next round starts along with the selection process of the set of cluster-heads.

In order to reduce intra-cluster interference, LEACH protocol proposes use of CDMA codes. Thus, when a node is selected as a cluster-head, it randomly selects a spreading code from a list. All nodes in the cluster are informed about this spreading code and they use this spreading code for further data transmission.

3.2.2.2 TEEN and APTEEN

TEEN or threshold sensitive energy-efficient sensor network protocol [16] is suitable for the applications for which sudden and drastic changes in the sensed parameters are to be noted. Thus, this protocol is well suited for time-critical applications.

The network model adopted for TEEN protocol is similar to the network model used for LEACH protocol. This model uses a hierarchical clustering scheme. It is assumed that the base station is far away from the nodes, but it has a continuous power supply and can transmit with high power to all the nodes. However, other nodes are energy constrained and direct communication from any node to the base station is expensive. Thus, clusters are formed with these nodes. A percentage of all the nodes are selected as cluster-heads and remaining nodes become associated with these cluster-heads forming clusters of nodes. Members in a cluster can communicate only with the corresponding cluster-head and the cluster-head aggregates the data and sends it to the base station. TEEN also supports second and higher levels in the hierarchy by allowing formation of cluster of clusters. One of the cluster-heads in a cluster of clusters is selected as the second level cluster-head and so on. Highest level cluster-heads report directly to the base station. The base station becomes the root of this hierarchy (Figure 3.6). Cluster-heads at higher levels in the hierarchy require to spending more energy as they transmit data over longer distances. Like LEACH, TEEN also supports rotation of cluster-heads so that energy consumption remains uniform across all nodes.

Thresholds

TEEN is implemented for applications that do not require the gathering of sensed data at periodic intervals, but rather require gathering of the data only when there is a drastic change in the values of the sensed parameters. Thus, TEEN uses thresholds to measure the data. However, instead of a single threshold value two different threshold values are used in this algorithm.

- Hard threshold (HT) is a threshold value for the sensed parameter beyond which the node sensing this value must transmit the data to its cluster-head.

- Soft threshold (ST) is the difference between the earlier sensed value and the current value of the same parameter. This value helps the node decide whether the current data will be transmitted to its cluster-head or not.

Functioning of the algorithm is the same as LEACH functioning during the set-up phase. A cluster-head, after being selected as the head, sends its members the values of the hard threshold and the soft threshold. When cluster-heads are changed, new values for the two parameters are broadcast again.

The nodes sense their environment continuously. Whenever sensed data exceeds the hard threshold value, the transmitter on the node is turned on

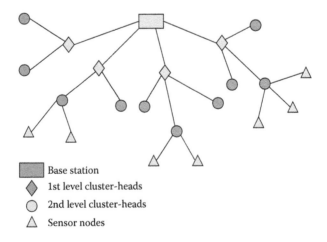

FIGURE 3.6
Multi-level cluster

and the data is transmitted to the cluster-head. The sensed value is stored in an internal variable, called the *sensed value* (SV). During the current cluster period (as long as the same set of cluster heads continues), the node again transmits data only when the following two conditions become true:

1. The current value of the sensed parameter is greater than the hard threshold.

2. The current value of the sensed parameter differs from SV by an amount equal to or greater than the soft threshold.

Thus, with the use of hard threshold, it is ensured that data will be transmitted only when its value falls in a range of interest for an application. Use of soft threshold further reduces the number of transmissions. This parameter ensures that data will not be transmitted if there is little or no change in the sensed parameter although its value is above the hard threshold. Therefore, a small value of the soft threshold is able to capture every minor change in the sensed parameter, thereby providing more accuracy to the application. On the other hand, this will also increase number of transmissions in the network. The trade-off between efficiency and data accuracy needs to be made at the application level.

APTEEN or adaptive periodic threshold-sensitive energy-efficient sensor network protocol [17] provides an improvement over TEEN by allowing the sensor nodes to send data periodically, and respond to sudden changes in the sensed parameters. In TEEN, the nodes never communicate if the sensed data do not exceed the threshold values. To overcome this drawback in TEEN, APTEEN proposes that during a cluster period, after cluster-heads are selected, they first broadcast the following four parameters:

1. Attributes (A): a set of parameters to be sensed by the nodes.

2. Thresholds consisting of a hard threshold (HT) and a soft threshold (ST). The uses of HT and ST are the same as in the case of TEEN.

3. Schedule: a TDMA schedule similar to the one used in LEACH assigning a slot to each node.

4. Count time (TC): the maximum time between two successive transmissions from a node to its cluster-head. This can be a multiple of the TDMA schedule length.

The algorithm works like TEEN. Nodes sense the environment continuously, and only when a node senses a data value beyond the hard threshold, it transmits the sensed data. Next, if its sensed parameter remains above the hard threshold, it transmits the data only when the change in the sensed parameter is equal to or greater than the soft threshold. However, the data sensed by a node may not exceed the hard threshold for a long time. In such cases, a node is forced to transmit its sensed data after a period equal to the count time (CT). The TDMA schedule compels each node in the cluster to transmit only during the assigned transmission slot.

APTEEN offers flexibility by allowing the user to set the CT and the threshold values. However, main drawback of this approach is the overhead associated with the implementation of threshold-based functions and CT in addition to setting up clusters at multiple levels.

3.3 Information Gathering Based on Geographic Locations

Many applications require the collection of data based on the geographic locations of sensor nodes or the location of a particular event needs to be spotted. For example, in case of forest fire detection, a natural query will be *"where is the fire?"* or in case of an intruder detection application, the location of the intruder should be precisely determined. Thus, location awareness of sensor nodes in such applications is important. Routing protocols are also implemented on the basis of this location information of the nodes.

However, adding location awareness to the sensor nodes is not a trivial task. One straightforward solution is to add *gobal positioning systems* or GPS devices to all nodes in the network. But the solution is not feasible due to the following reasons:

- GPS devices are costly and when the sensor nodes must be deployed in large numbers, use of GPS devices with every node will make the deployment expensive and impractical.

- Power consumption by GPS devices is high and these devices cannot be used with energy-constrained nodes where continuous power sources are not available.

- Due to the size of GPS devices and their antennas, the form factor of sensor nodes increases, thereby making them unsuitable for field deployment.

- GPS devices are unable to work in environments where obstructions block the line-of-sight communications from the satellites. Thus, they cannot work in indoor environments, in forests or in places where obstacles are present.

Therefore, localization techniques must be used to make the sensor nodes location-aware to implement geographic routing. In the next subsection important localization techniques are discussed and geographic routing algorithms are discussed in Section 3.3.2.

3.3.1 Localization

Major techniques for localization of sensor nodes (making the sensor nodes location-aware) depend on use of anchor nodes and estimation of distances (or angles) locally. Anchor nodes know their own physical positions and sometimes positions of other nodes. For some applications, logical locations, i.e., coordinate systems or symbolic references, are also used. Nodes which are either directly adjacent or at multiple-hop distance from the anchor nodes can be localized by estimating and combining their distances (or angles) with respect to the anchor nodes.

3.3.1.1 Localization Basics

Distance Estimation

Some popular techniques for estimating distances between two nodes are discussed in this section [24].

Received Signal Strength Indicator (RSSI)

A signal of known strength is sent out. Power of the signal at the receiver end is measured and compared with the power of the transmitted signal which is known beforehand. Effective propagation loss is calculated and theoretical and empirical models are used to translate this loss into a distance estimate between the sending and the receiving nodes. RSSI is useful for RF signals. However, the technique is highly error-prone and is applicable only for short ranges.

Time-based Methods

Time-based approaches for distance estimation depend on the signal propagation speed and time required for propagation of a signal from one node to another. Two approaches are implemented.

- **ToA** or time-of-arrival of a signal is recorded by noting the time when transmission starts and the time when the signal reaches the receiving node. The propagation time is then directly translated into distance based on the known signal propagation speed. To implement this technique, exact time synchronization is needed.

- **TDoA** or time-difference-of-arrival involves two different signals with different propagation speeds. For example, an ultrasound signal and a radio signal can be transmitted from a node. Propagation time of a radio signal is negligible compared to an ultrasound signal. The difference between their arrival times at the receiving node is computed and translated to distance based on the propagation speeds of the signals. This technique requires expensive and energy-intensive hardware and precise calibration.

The above techniques can be applied to many different signals such as RF, acoustic, infrared and ultrasound.

Angle-of-Arrival (AoA)

Signals are transmitted from a sender node and the angles at which signals are received at the receiving node are measured. Geometric relationships are then used to calculate node positions.

Distance Combining

After distance estimation, the next phase is *distance combining*. Some of the techniques in the combining phase are discussed below.

- **Hyperbolic trilateration** - Location of a node is estimated by calculating the intersection of three circles. Trilateration is a three-step process. Initially, the node tries to find its location with respect to one of the anchor nodes (also referred to as *beacon nodes*). In this phase, the node estimates its position to be loci of points on the circumference of a circle with the anchor node at its centre. Radius of the circle is distance between the node and the anchor node estimated in the *estimation phase*. In the next phases, the node finds its location with respect to two other anchor nodes in similar manner. Finally, location of the node is computed as the intersecting point of the circles formed with respect to all three anchor nodes. Since the intersection of all three circles can be one single point, location of a node can be computed in straightforward manner with reasonable accuracy.

- **Triangulation** - The technique is used when the directions of the node with respect to anchor nodes are estimated instead of its distance from the anchor node. The AoA system is applied for distance estimation. The node positions are calculated using trigonometric rules.

- **Maximum likelihood (ML) estimation** - It estimates the position of a node by minimizing the differences between the measured distances and estimated distances.

- **Atomic multilateration** - This method is used to find the location of a node when at least three beacon nodes with known locations are available at one-hop distance. Suppose the node whose location is to be decided is denoted as S and its position is given by (x_0, y_0). The position of every beacon node is known. Consider that the i-th beacon node is located at (x_i, y_i) for $i = 1, 2, 3, \ldots N$, where N is the number of beacon nodes at one-hop distance from S. Figure 3.7(a) shows that three beacon nodes, P, Q and R, are available at one-hop distance from S. Considering s to be the estimated propagation speed of an ultrasound signal and time taken for the signal to propagate from the i-th beacon to node S is t_{i0}, a set of equations are formed to express the difference between the measured distance and the Euclidian distance with respect to each beacon node.

$$f_i(x_0, y_0, s) = st_{i0} - \sqrt{(x_i - x_0)^2 + (y_i - y_0)^2}.$$

A unique solution to (x_0, y_0) can be obtained using the above equations when at least three beacon nodes are available in the vicinity of S. If four or more beacon nodes are available, the speed of ultrasound signal can also be estimated.

- **Iterative multilateration** - Iterative multilateration is implemented to find the locations of a set of nodes with respect to a few beacon nodes. First a node (node S in Figure 3.7(a)) with unknown location and maximum number of beacon nodes in its neighbourhood is considered. The location of S is estimated using *atomic multilateration*. Once the location of the node is known, it becomes a beacon node and is used for finding the locations of other nodes. For example, in Figure 3.7(a), node T initially has only two beacons but after the location of S is known, it can be used as the third beacon node.

The major drawback of this scheme is the error accumulation due to the use of a node with an initially unknown location (whose location is estimated using atomic multilateration) as a beacon node in a later stage. If there is any error in location estimation in the first place, the error will also influence the location estimation of subsequent nodes.

- **Collaborative multilateration** - In a random deployment of sensor nodes, it is possible that some nodes may never have three neighboring beacon nodes and therefore, the conditions for atomic multilateration cannot be met. Figure 3.7(b) shows one such example topology. Nodes P, Q, R and S are beacon nodes and the locations of nodes A and B are to be determined. When such a situation occurs, an attempt is made to estimate the positions of nodes by considering location information over multiple hops. This technique is known as *collaborative multilateration*. The two requirements for collaborative multilateration are:

  1. A node is a participating node if it is either a beacon or a node with unknown location having three neighboring participating nodes.
  2. A participating node pair is a beacon-unknown node pair or an unknown-unknown pair in which all unknowns are participating.

A set of simultaneous quadratic equations are formed on the basis of information available about the participating nodes and if sufficient information is available, the set of equations is solved to obtain a unique solution set that estimates the positions of the nodes with unknown locations.

Localization Schemes

Based on the above techniques, a number of localization schemes have been proposed. The schemes either adopt a centralized approach or a distributed approach.

Centralized algorithms are designed to run on a node with high computational power and ample resources. The base station can be used to run

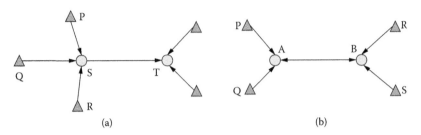

FIGURE 3.7
Multilateration: (a) iterative multilateration, (b) collaborative multilateration

centralized algorithms. Sensor nodes gather environmental data and pass it back to a base station for computation. The computed positions are transmitted back to the nodes in the network. Although centralized computation can solve the localization problem more accurately since the central node possesses all required information, the necessity to send and receive information to a central point creates bottlenecks in the system and power consumption is high due to more traffic. Time synchronization is another problem. The latencies for location updates are high because of central computation.

On the other hand, if computation is distributed among the nodes and performed locally, no time synchronization among the nodes is required. The schemes can be implemented using less traffic and location information can be updated rapidly. Distributed computation is more robust to node failure.

There are two approaches to distributed localization. A set of nodes can implement beacon-based distributed algorithms when beacons are deployed in the network. Nodes in the network use one of the above-mentioned distance combining techniques. The second approach focuses on optimization of a global metric over the network in a distributed fashion.

Some centralized and distributed localization schemes are discussed in the following subsections.

3.3.1.2   Centralized Algorithms

This section discusses some example centralized algorithms to demonstrate their functioning.

1. **Semidefinite programming (SDP)** - In this algorithm [7], geometric constraints between nodes are represented as linear matrix inequalities (LMIs). In general, only constraints that form convex regions can be used for representation as LMIs. Thus, angle of arrival data can be represented as a triangle and hop count data can be represented as a circle. Once all the constraints in the network are expressed in this form, the LMIs can be combined to form a single semidefinite program. This is solved to produce a bounding region

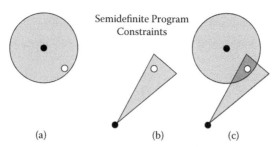

FIGURE 3.8
Geometric constraints of a node: (a) radial constraint, for example from radio connectivity, (b) triangular constraint, for example from angle of arrival data, (c) location estimate derived from intersection of two convex constraints [3]

for each node. For example, a radial constraint can be obtained from radio connectivity of a node and a triangular constraint can be obtained from angle of arrival data. Next, location estimate of a node can be derived from intersection of these two convex constraints. Because of the computational complexity of solving the semidefinite program, this algorithm must be run on a central node. However, this makes the algorithm non-scalable. Also all constraints cannot be expressed as LMIs. Therefore, the algorithm is not suitable for practical use.

2. **MDS-MAP** [25] uses a technique called multidimensional scaling (MDS). The basic logic behind this concept is that if there are n points suspended in a volume whose locations are unknown and only distances between each pair of points are known, the laws of cosine and linear algebra can be used to reconstruct the relative positions of the points.

   The algorithm works in four stages:

   Step 1: A sparse matrix is formed with ranging data (distance information) for each pair of nodes. An entry r_{ij} in matrix R is zero if there is no connectivity between node i and node j; otherwise it is the distance between two nodes.

   Step 2: A standard all-pairs shortest path algorithm (e.g., Dijkstras or Floyds algorithm) is used to produce a complete matrix of inter-node distances D.

   Step 3: MDS is applied on D to find estimated node positions X.

   Step 4: X is transformed into global coordinates using some number of fixed beacon nodes following a coordinate system registration routine [3].

3.3.1.3   Beacon-based Distributed Algorithms

This section presents some beacon-based distributed algorithms as examples.

1. **AHLoS** - The *ad hoc localization system* [24] is based on fine-grained point-to-point distance estimation using time-of-arrival or TOA. Beacon nodes are used for estimating the locations of non-beacon nodes. The scheme implements iterative multilateration and collaborative multilateration for estimating node positions. Though the scheme works well for small networks, it fails to provide accurate estimates for large networks due to accumulation of errors.

2. **APS** - The *ad hoc positioning system* [20, 21] enables hop-by-hop propagation of distances to estimate the distance of any arbitrary node from a beacon node or landmark (a node taken as a reference is termed a *landmark*; APS assumes that the landmarks are GPS enabled, and therefore, act as beacon nodes). A node may not be within the range of a beacon node, but its distance from a beacon node is estimated in a hop-by-hop manner. The immediate neighbors of a landmark can estimate the distance to the landmark by direct RSSI or other techniques. Using one of the propagation methods discussed below, the second-hop neighbors then infer their distances to the landmark. Remaining nodes in the network continues inferring their distances in similar manner. Thus, distance information flows in a controlled flood manner, initiated at the landmark. Once an arbitrary node obtains its estimated distances to a number of landmarks (greater than or equal to three), it can compute its own position in the plane. Thus, each node only communicates with its immediate neighbors, and during every message exchange the estimated distances of a node to landmarks are communicated. One of the following three hop-to-hop distance propagation methods can be used:

   DV-hop propagation method - First a classical distance vector exchange is used, so that all nodes in the network know their distances in hops to the landmarks. Initially, the hop distance is assumed to be the maximum transmission range of a node. Once a landmark knows its distance in hops from other landmarks, it estimates an average size for one hop by dividing the Euclidian distances by number of hops. This average size of a hop is then used as a correction to the entire network. After receiving the correction, an arbitrary node may compute estimated distances to landmarks in meters.

   For example, in Figure 3.9, L_1, L_2, L_3 are landmarks. Euclidian distances of L_1 from L_2 and L_3 are 40 m and 100 m respectively and hop distances are 2 hops and 5 hops. L_1 computes a correction $(100 + 40)/(2 + 5) = 20$ as the average hop distance.

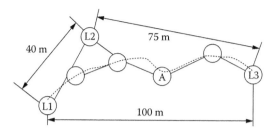

FIGURE 3.9
DV-hop propagation method

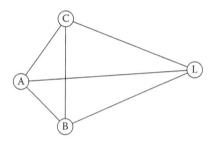

FIGURE 3.10
Euclidian propagation method

Similarly, L_2 computes a correction 19.17 and L_3 computes a correction 19.44. These corrections may be broadcast in the network and a node gets the correction from its nearest landmark and applies the correction to estimate its distances from landmarks and from other nodes. Here A gets its correction from $L2$ which is its nearest landmark.

DV-distance propagation method - This is similar to the previous method except the distance between two neighboring nodes is measured using RSSI or other techniques and is propagated in meters rather than in hops. Because all hops may not have the same length, this method works better than the previous one. However, use of RSSI can always influence the accuracy of measurements.

Euclidean propagation - This method propagates the *Euclidian* distance to the landmarks. An arbitrary node A with at least two neighbors B and C and a landmark L forms a quadrilateral $ABCL$ as shown in Figure 3.10. Distances of B and C from L are known and distances of A from B and C can be measured using distance estimation techniques. These distances provide an estimate of AL.

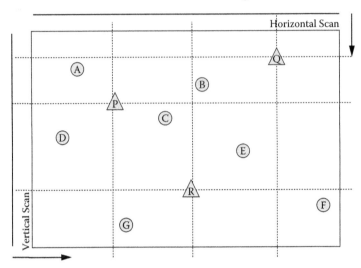

FIGURE 3.11

Obtaining node sequences [27]

3. **MSP** - The functioning of this algorithm (multi-sequence position-
    ing of wireless sensor nodes) [27] depends on extraction of relative
    location information from multiple simple one-dimensional order-
    ings of nodes. MSP considers a sensor network with few anchor or
    beacon nodes and many target nodes whose locations are to be de-
    termined. First some events are generated, one at a time, at different
    locations within the network. Examples of such events are propaga-
    tion of ultrasound signals from different locations or laser scans with
    diverse angles. Each node in the network detects the event at some
    instance of time. For a single event, the nodes are ordered based on
    the sequential detection of the event. A node sequence includes both
    anchor nodes and target nodes. Next, a multi-sequence processing
    algorithm is used to split the entire sensor network area into small
    parts that become the possible locations of each node. Finally, a
    distribution-based method is used to estimate the location of each
    sensor node.

    Figure 3.11 explains the partitioning of the network using two
    straight line scans. Anchor nodes are marked with triangles and tar-
    get nodes are marked with circles. Vertical scanning generates the
    node sequence (D,A,P,G,C,R,B,E,Q,F), and horizontal scan gener-
    ates the node sequence (Q,A,B,P,C,D,E,R,F,G). Since the locations
    of the anchor nodes are available, the anchor nodes in the two node
    sequences split the area into 16 parts, as shown in Figure 3.11. Pre-
    decessor and successor anchors of each target node in a sequence

determine the boundaries of the node. Every time a new boundary is obtained by a new event, the location area of the node shrinks. After processing all sequences, a final polygon is obtained whose centroid becomes the estimated location of the target node. Thus, the accuracy of location estimation depends on the number of anchor nodes and also on the number of events occurred in the network.

Further improvements of the basic MSP as described above have been suggested [27]. Advanced MSP makes use of the full information embedded in a sequence and narrows the small area for each node.

While, AHLoS is a range-based algorithm, i.e., it depends on distance or angle measurements, APS is range-free and does not need to estimate accurate distances or angles. Location of each node is estimated based on the known locations of the anchor nodes. Different combinations of estimated distances from anchor nodes narrow the areas in which the target nodes can possibly be located. Both AHLoS and APS require high densities of anchor nodes. On the other hand, MSP attempts to reduce the number of anchor nodes and relies more on event distribution.

3.3.1.4 Beacon-free Distributed Algorithms

Moore et al. [18] present a distributed localization algorithm that works without anchor nodes and localizes nodes in regions constructed from robust quadrilaterals. The idea is the same as for the quadrilaterals used in *Euclidian propagation* in *APS*. It has been shown that a quadrilateral is the smallest possible subgraph that can be unambiguously localized. Consider the four-node subgraph in Figure 3.12(a). It is fully connected and there are six edges whose distances can be measured. The following properties of the quadrilateral become useful:

1. Relative positions of the four nodes are unique up to a global rotation, translation, and reflection. In graph theory terms, the quadrilateral is globally rigid.

2. Any two globally rigid quadrilaterals sharing three vertices from a five-vertex subgraph are also globally rigid. By induction, any number of quadrilaterals chained in this manner form a globally rigid graph.

A quadrilateral can be decomposed into four triangles as shown in Figure 3.12(b). If the smallest angle in one of these triangles is close to zero, there is a chance of measurement error. Thus, a robust quadrilateral must only contain triangles with sufficiently large minimum angles. Thus, Moore et al. [18] define the following property which a robust triangle must satisfy

$$b \sin^2 \theta > d_{min}$$

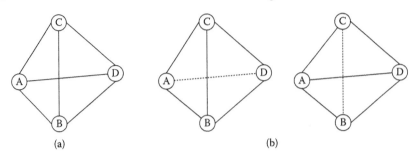

FIGURE 3.12
Robust quadrilaterals: (a) robust four-vertex quadrilateral, (b) decomposition of robust quadrilateral into four triangles

where b is the length of the shortest side and θ is the smallest angle. The threshold value d_{min} is chosen based on the measurement noise. A robust quadrilateral is defined as a fully-connected quadrilateral whose four sub-triangles are robust.

The algorithm proposed in [18] works in three phases:

Cluster localization - A cluster is defined as a node and its neighbors. Thus, each node becomes the center of a cluster and estimates the relative location of its neighbors. For each cluster, all robust quadrilaterals are identified and the largest subgraph consisting of overlapping robust quadrilaterals is considered. Locations of nodes within the cluster can be estimated incrementally by following the chain of quadrilaterals and trilaterating along the way.

An example is shown is Figure 3.13. Nodes A, B, C, D form a robust quadrilateral with A as the origin of a local coordinate system. Positions of B, C, and D are determined that satisfy the six distance constraints given by four sides and two diagonals of the quadrilateral. In the next step, the second robust quadrilateral $ABDE$ is chosen and node E is localized relative to the

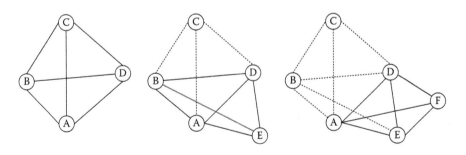

FIGURE 3.13
Cluster localization with overlapping robust quadrilaterals

known positions of A, B, and D. A trilateration technique can be used for this. Continuing with the same procedure, node F is localized which is part of the robust quadrilateral $BDFE$.

Cluster optimization - This is an optional phase. The position estimates for each cluster are refined using numerical optimization such as spring relaxation or Newton-Raphson with the full set of measured distance constraints. Any error that accumulates during the incremental approach adopted in the first phase can be reduced using this approach. No communication overhead is required during this phase because the process is performed per cluster and not for the network as a whole.

Cluster transformation - The local coordinate systems of neighboring clusters are transformed by finding the set of nodes in common between two clusters and bringing the nodes of the two local coordinate systems into best coincidence by applying rotation, translation, and possible reflection that best align the clusters.

A node that is not part of the largest subgraph of robust quadrilaterals in the cluster cannot be localized using the above algorithm.

The problem with anchor-free localization is that same distances between pairs of nodes can be realized differently. Thus, the nodes can move continuously without violating the distance constraints or there is possibility of *flip ambiguity* which completely changes the estimated position of a node in comparison with its actual position. Rigidity theory of graphs has been used by many researchers to overcome these problems. A lot of research works target measuring and analyzing localization errors and minimizing them.

A localization algorithm needs to be selected in such a way that it can scale for indoors, outdoors or global scenarios based on the application requirements. Accuracy (how close an estimated position is to the real position) and precision (for repeated position determinations, how often a given accuracy is achieved), cost of implementing the algorithm, energy consumption and other issues must also be considered while choosing options.

3.3.2 Geographical Routing

Geographical routing is based on the assumption that the nodes know their geographical locations and also know the locations of their one-hop neighbors. In case of data-centric routing algorithms discussed in Sections 3.2.1 and 3.2.2, all data packets are forwarded to the sink. On the contrary, in the case of geographic routing, routing destinations are specified geographically, i.e., either a location, or a geographical region is mentioned while initiating data transmission. The source node is aware of the geographic location of the destination. Each packet holds routing information along with data. A node forwards a data packet to the one-hop neighbor that makes progress towards the destination. Progress can be measured as Euclidean distance of the destination node or as the angle towards the destination (the smallest angle is chosen), which is termed *compass routing*. No flooding is involved in geographical routing. Connectivity graph among the nodes is modeled as a unit disk graph.[1]

In this section some geographic routing protocols are discussed.

3.3.2.1   Greedy Perimeter Stateless Routing

Greedy perimeter stateless routing (GPSR) [2] considers a topology where the wireless sensor nodes are roughly in a plane. GPSR also works on other ad hoc networks including IP-based networks. Each node only knows the positions of its next-hop neighbors and the position of the destination node. These positions are sufficient to make correct forwarding decisions without any other topological information.

In GPSR, the source node generates packets which contain the location of the destination. A node that needs to forward a packet makes a greedy choice to determine the next hop. In particular, it selects the next-hop neighbor which is geographically closest to the destination. The next-hop neighbor, upon receiving the packet, makes a greedy choice to forward it closer to the destination The process continues until the destination is reached.

Each node periodically broadcasts a beacon containing only its own identifier (e.g., IP address or other identifier) and its position. The position of a node is encoded as two four-byte floating point quantities for x and y coordinate values. A table is maintained by each node to store the positions of its neighbor nodes. If a node does not receive a beacon from a neighbor node for a long time (longer than a time-out interval T defined earlier), the GPSR node assumes that the neighbor has failed or gone out of range (in cases of mobile nodes), and deletes the neighbor from its table.

In order to minimize the cost, position of a node can be piggybacked on all data packets forwarded by it. All nodes within the radio range can receive the packet and know the position of the sending node. Although some additional bytes are added with the data packet to hold this information, the piggybacking scheme reduces beacon traffic in regions of the network where data packets are forwarded frequently.

The other proposal is that instead of broadcasting beacons at regular intervals, a node can solicit beacons by broadcasting a neighbor request message only when it wants to forward data packets. Any of the schemes can be chosen based on the application requirements.

Problems with Greedy Choices

Greedy choices can sometimes lead to failure if, in spite of the existence of a route to the destination, a packet is required to move temporarily farther from the destination. This situation may occur specifically when there is a

---

[1]A unit disk graph is the intersection graph of a family of unit disks in the Euclidean plane. That is, it is a graph with one vertex for each disk and an edge between two vertices whenever the corresponding disks have non-empty intersections.

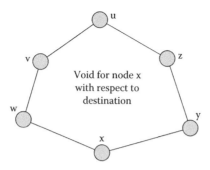

FIGURE 3.14
Problems with greedy choices: node x cannot send directly to destination u; it requires forwarding of packet to distant node y

hole in the network as shown in Figure 3.14. In such cases, GPSR forwards the packets following the perimeter of the planar graph to find the route.

Combined Algorithm Based on Greedy and Planar Perimeters

A graph in which no two edges cross is known as a planar graph. It subdivides the plane into connected regions called faces. A set of homogeneous nodes with circular transmission range can be seen as a graph with each node as a vertex. An edge exists between nodes n and m if they are within the radio range of each other. It has been shown [2] that a planar graph can be obtained from such a graph without losing the connectivity.

All nodes in the network maintain a neighbor table which stores the addresses and locations of their single-hop neighbors. When perimeter-mode forwarding is adopted, the packet header fields used in GPSR include location of the destination D, location of the packet that entered into perimeter mode L_p, point where the packet entered the current face of a planar graph L_f, first edge traversed on current face e_0 and the packet mode (*greedy or perimeter*) M.

Destination location of a packet is set by a source node only. Initially all packets are forwarded in greedy mode. Upon receiving a greedy-mode packet for forwarding, a node searches its neighbor table for a neighbor closest to the destination. If no such neighbor is found, the packet is marked as a perimeter mode packet. The node where a packet enters into perimeter mode is marked as x and a straight line $\overline{xD}$ is considered between x and and destination D. In perimeter mode, a packet is forwarded on progressively closer faces of the planar graph, each of which is crossed by the line $\overline{xD}$. The face can be an *interior* (closed polygonal region bounded by edges) or an *exterior* (unbounded face) face. While traversing the planar graph, the right hand rule is always adopted. When an edge is reached that crosses the line $\overline{xD}$, the algorithm moves to the adjacent face crossed by $\overline{xD}$. In other words, it keeps left on

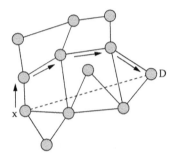

FIGURE 3.15
Perimeter forwarding

the boundary of a face, proceeds until it hits the straight line connecting the node x to destination, then it switches to the next face. Figure 3.15 shows an example of traversal.

The packet forwarding returns to greedy mode only when the distance between the current forwarding node and the destination is less than the node where the algorithm entered into perimeter mode.

In case the destination is disconnected, data forwarding will return back to the node where it enters the perimeter mode and fail as there is no path to the destination. Otherwise, delivery of a message is guaranteed if there is a path between source and destination.

3.3.2.2   Geographical Energy Aware Routing

Geographic and energy aware routing (GEAR) [26] is based on the *Directed Diffusion* algorithm. The key idea is to restrict the number of interests in Directed Diffusion by only considering a certain region rather than sending the interests to the whole network. Thus, GEAR uses a heuristic for geographically informed neighbor selection.

Typically, in any network, packets are forwarded to a destination node (or a set of nodes). In contrast, GEAR considers a target region instead of a particular node. Thus, it suits data-centric sensor networks. The protocol assumes that each node knows its location using a GPS device or any other means as discussed in Section 3.3.1. Each node is also informed about its remaining energy level and the locations and energy levels of its neighbors (a beacon exchange protocol similar to GPSR can be implemented for this). GEAR also assumes that the sensor nodes are static.

In the first phase, a packet is forwarded towards the target region based on a greedy choice. However, unlike GPSR, the choice of neighbor is based on the location of the neighbor node and the node energy level. If there is a neighbor closer to the destination region than the current node, it is selected as next-hop neighbor. In case there is a hole or energy level of the neighboring node is low, GEAR choses another neighboring node based on *learned cost calculation*. GEAR will be discussed in detail in Chapter 4.

3.3.2.3 Face Routing Protocols

Kuhn et al. present a geometric ad hoc routing protocol known as GOAFR or greedy other adaptive face routing in [14]. This algorithm primarily focuses on ad hoc networks. GOAFR considers a unit disk graph (UDG) network model. Nodes in a UDG are positioned on a Euclidean plane and two nodes are connected only if distance between them is ≤ 1. Thus, UDG models a flat environment with nodes equipped with wireless radio, all having equal transmission ranges. GOAFR is an extension of the basic face routing or FR algorithm. A number of variations of face routing protocols have been proposed as described in this section.

Face Routing (FR)

FR explores the boundaries of faces in a planar graph using the local right hand rule (in analogy to following the right hand wall in a maze). On its traversal, it keeps track of each intersection point with the line connecting the source to the destination. After completely traversing a face, the algorithm returns to one of the intersection points closest to the destination and then moves to the next face.

BFR and AFR

The main disadvantage of FR is that it requires exploration of all the face boundaries. *Bounded face routing* or BFR restricts this exploration of face boundaries by restricting the searchable area. The area is defined by an ellipse whose size is equal to the length of the optimal path connecting the source and the destination. However, it may not be possible to find the length of an optimal path. Thus, *adaptive face routing* or AFR extends BFR by starting with the ellipse size equal to an initial estimate of the optimal path length. If the destination is not reached, BFR is restarted with a bounding ellipse of double size.

OFR and OAFR

Other face routing (OFR) differs from face routing in that the starting point of exploring the next face is different. Thus, instead of starting exploring the next face from an intersection point on the line connecting the source and the destination, OFR starts exploration from a boundary point on the intersection of the current face and the next face closest to the destination. Figure 3.16 shows the two cases; s is the starting node and t is the destination node. FR starts at s, and traverses through the face F_1. It finds P_1 on the line joining s and t and moves to F_2. OFR finds P_3, a point on the boundary of F_1 as the closest point to the destination. Therefore, it moves to F_4.

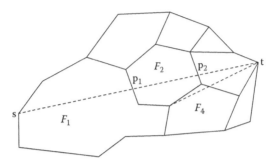

FIGURE 3.16
FR and OFR

Other Adaptive Face Routing or OAFR is the adaptive routing using OFR.

GOAFR

Finally, *greedy other adaptive face routing* or GOAFR combines the greedy approach with OAFR. Thus, the idea is similar to GPSR. GOAFR performs greedy steps until it reaches the destination or a local minimum is met. At the time of traversal, if the next step leads beyond the elliptical boundary, the search area is increased by doubling the length of the major axis of the ellipse. Local minimum is met when a node has no neighboring node which is closer to the destination. Then the algorithm executes OAFR on the first face only, doubles the length of the major axis of the elliptical boundary and repeats the step.

It has been claimed that the algorithm is optimal in the worst case and performs efficiently on average. The asymptotic optimality of GOAFR is shown in References [13] and [14].

3.3.2.4   Modified SPIN

A *modified SPIN* or M-SPIN protocol has been introduced [23]. This protocol has been designed to transmit information only to sink node instead of transmitting throughout the network. The protocol is based on the SPIN family of protocols.

In M-SPIN protocol, a new phase called *distance discovery* is added to find the distance of each sensor node in the network from the sink node in terms of hops. This means that nodes having higher values of hop distance are far away from the sink node. Other phases of M-SPIN are *negotiation* and *data transmission*. On the basis of hop distance, negotiation is done for sending actual data. Therefore, the use of hop values controls dissemination of data in the network. Finally, data is transmitted to the sink node.

Distance Discovery

Initially the sink node broadcasts a startup packet in the network with type, nodeId and hop. The type field stores the types of messages, nodeId stores the identification of the sending node and hop represents hop distance from the sink node. Initial value of the hop is set to 1. When a sensor node receives the startup packet, it stores this hop value as its hop distance from the sink node in memory. After storing the value, the sensor node increases the hop value by 1 and then re-broadcasts the startup packet to its neighbor nodes with the modified hop value. It is possible that a sensor node receives multiple startup packets from different intermediate nodes. Whenever a sensor node b receives startup packets from its neighbors $a_i, 1 \leq i \leq n$, it checks the hop distances and sets the distance to the minimum.

The process continues until all nodes in the network get the startup packets at least once within the the *distance discovery* phase. After successful completion of this phase, the negotiation phase starts.

Negotiation

The negotiation phase is similar to the SPIN-BC protocol with an important difference. The source node sends an ADV message. Upon receiving an ADV message, each neighbor node verifies whether it has already received or requested the advertised data. Unlike the usual SPIN algorithm, a receiver node also verifies whether it is closer to the sink node or not in comparison with the node that sent the ADV message. This is the main difference between the negotiation phases of SPIN-BC and M-SPIN. If the hop distance of a receiving node is less than the hop distance of the sender of the ADV message (this value is received as part of the ADV message), the receiving node sends REQ message to the sending node for current data. The sender node then sends the data to all the nodes that requested the data, using DATA message.

As soon as a node gets data from its own application or from other sensor nodes, it stores the data in its memory. When an ADV message is received, each receiving node first checks its record to ascertain whether it already has seen that data. Similarly, when a requesting node receives data, it immediately checks whether the data is the same for which it has sent the request.

Data Transmission

Data transmission is the same as the SPIN-BC protocol. After a request is received by the source node, data is immediately sent to the requesting node. If the requesting nodes are intermediate nodes other than the sink node, the *negotiation* phase repeats. Thus, the intermediate sensor nodes broadcast ADV for the data with modified hop distance value. The sending node modifies the hop distance field with its own hop distance value and adds that in packet format of the ADV message. The process continues until data reaches the sink node.

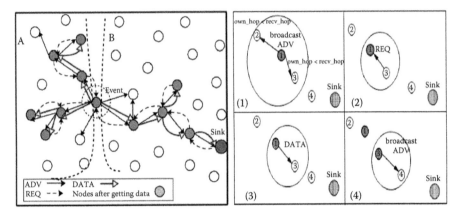

FIGURE 3.17 (SEE COLOR INSERT)
M-SPIN protocol in WSN: (a) partitioning into two regions, (b) negotiation using M-SPIN

In Figure 3.17(a), when an event occurs in the network, it divides the entire network into regions A and B. Sensor nodes in region A are on the other side in the network in comparison with the sink node and sensor nodes in region B are on the same side and nearer to the sink node. Sensor nodes of region A can receive data from the node sensing the event; however, they will unnecessarily waste their energy in receiving or transmitting the data. In order to deliver data to the sink node, data will have to travel more hops if nodes in region A participate in data forwarding. Thus, when an event occurs, it is always desirable that the data is sent through the nodes in region B.

Figure 3.17(b) depicts the functioning of the M-SPIN protocol.

3.3.3 Landmark-based Routing

Algorithms like GPSR depend on actual coordinates or geographic locations of the nodes. Though they are effective for a dense network on a two-dimensional plane, it is difficult to implement these algorithms in the presence of obstacles within the network. In such cases, the positions of two nodes may be quite close, but direct communication between them may not be possible. Further, if the actual deployment is in three dimensions, forcing a two-dimensional layout (to apply planarization) can create distortion. In some research works [19, 22], authors proposed use of virtual node coordinates and geographic routing protocols, like GPSR, on these virtual coordinates. These are obtained by embedding the link connectivity graph of the nodes in a plane. Thus, nodes that can communicate directly are embedded close to each other and those do not have communication links are embedded further away. Virtual coordinates-based routing is also discussed [5, 6, 9].

Encouraged with the use of virtual coordinate systems, some researchers proposed landmark-based routing. The key idea behind landmark-based routing is as follows. In the network, k nodes are selected as landmarks or beacons which flood the network. Each node records hop distances to these landmarks. This distance measurement enables point-to-point routing in the networks. These routing schemes are simple, location-free and work in three-dimensional network. No unit disk assumption is required. GLIDER is an example of such routing discussed in the next subsection.

3.3.3.1 Gradient Landmark-based Distributed Routing for Sensor Networks

Gradient landmark-based distributed routing for sensor networks [8] or GLIDER primarily depends on node connectivity and not on any knowledge of node positions. Two major phases are executed in this algorithm: a global preprocessing step and a local step to solve local routing problem.

In the preprocessing step, the global topology of the sensor network is constructed, and provides information about the connected components, holes or obstacles (e.g., information about the buildings within the network area) in the sensor network and general routing guidance (for moving around a building or turning left). During this process, the entire network field is partitioned into tiles. Next a greedy forwarding method based on local coordinates is implemented within each tile to capture the local connectivity and gradient descent to realize the routing path. Implementation of this two-level infrastructure is useful because the global topology of a network is generally stable (the positions of buildings are unlikely to change often) and the global topology is compact (a small number of buildings). When a rough guidance is obtained, local greedy rules are sufficient for further routing of data packets.

The GLIDER algorithm starts with a communication graph of sensor nodes. The edges are unweighted and only identify which pairs of nodes have direct communication (not the geometric distances between the nodes). Distance between two nodes is given by number of edges (i.e., hop count) in the shortest path between the two nodes. After obtaining the communication graph, a set of landmarks is selected among the nodes. These landmarks flood the network with beacons and each node learns its distance in hop count to each landmark. The algorithm then proceeds to construct a *landmark Voronoi complex* or (LVC).

Landmark Voronoi Complex

Landmark Voronoi is the natural extension of the geometric Voronoi diagram and similarly, a Delaunay complex can be obtained from its dual Delaunay triangulation. To construct an LVC, each sensor identifies its closest landmark. A sensor may have same distance from two or more landmarks; then the node is placed on the boundary. Construction of the LVC partitions the network area. As the spanning tree of each Voronoi cell is connected, the Voronoi cells of a set of landmarks provide a natural partitioning of the sensor network area into connected tiles.

Based on the Voronoi cells belonging to the landmark Voronoi complex (LVC), combinatorial Delaunay triangulation (CDT) is used to record the adjacency relation between the Voronoi cells. If there is at least one boundary node between landmarks i and j, there is an edge ij in CDT. Holes in the network area become holes in the CDT. It can be proved that if G is connected, then the combinatorial Delaunay graph $D(L)$ for any subset of landmarks is also connected (interested readers may refer to [8] for details). The CDT is broadcast to the entire network to give a global topology. $D(L)$ can be used to provide global connectivity information which is accessible to every node for proactive route planning on tiles.

Landmark Selection

Selection of landmarks is important for effective execution of the algorithm. Since, CDT provides the global guidance and is stored by every node, it should be small and simple. On the other hand, sufficient landmarks are needed to ensure that each Voronoi cell has a simple topology (e.g., without any hole or obstacle). These two goals are contradictory to each other. Thus, landmarks should be selected to capture all topological features.

Local Landmark Routing

Each node stores the shortest path tree on D(L) rooted at its home landmark. A node also stores its own coordinates and the coordinates of its neighbors for greedy routing. Here coordinates are only virtual, i.e., $(d_1, d_2, d_3, ...d_k)$, where d_k = hop count to the k-th reference landmark (home + neighboring landmarks). With the virtual coordinates, a node can check if whether it is on the boundary (i.e., two landmarks are at the same distance). It can also find a neighbor who is closer to a reference landmark.

Local routing scheme handles two issues: how to move from tile to tile (*inter-tile routing*) and also routing within the tile (*intra-tile routing*). High-level routes on tiles are realized as actual paths in the network by using reactive protocols. Intra-tile routing is done by gradient descent using the local landmark coordinates. When a packet reaches a node, it is relayed to a neighbor which is closer to the landmark based on local landmark coordinates.

3.4   Data Aggregation

In a sensor network, data are collected from the sensor nodes in the following three manners:

1. Periodical measurement: Sensor nodes periodically send reports to the sink or base station.

2. Query-based sensing: The sink or base station either floods or sends queries to the sensor nodes in specific regions for currently sensed information.

3. Event detection: Occurrence of a certain event can trigger the sensor nodes in that region to forward the detected event in a proactive manner.

In the case of a query-based approach, a query can be data-centric ("find a four-legged animal"), or geographical ("get the temperature of the north-east corner of the room"), or multi-dimensional (involving spatial, temporal and other types of attributes or attribute range).

However, since sensor nodes in a dense network often detect common phenomena, there may be redundant data from various sources communicating to a particular sink. A large volume of data can also be reduced through application of in-network filtering and processing techniques.

In-network processing, that is, processing data at the time of routing, is one of the major requirements of wireless sensor networks. As mentioned, this reduces the network traffic and conserves energy of the sensor nodes. Other reasons for in-network processing are to discover some hidden patterns or generate a global picture, which cannot be obtained from a single sensor data. Sensor data from multiple sources also reduces the chance of erroneous data and uncertainty in data and outliers.

Data aggregation involves fusion of data from multiple sensors at intermediate nodes and transmission of the data to the base station or sink. The idea is to combine the data from different sources to eliminate redundancy and minimize the number of transmissions.

Three questions need to be addressed in a sensor network. First, at what instance does a node report a sensed event? Second, how does a node fuse multiple sensed data into a single one? Third, what data aggregation architecture should be used?

Data aggregation algorithms can be implemented on a centralized, distributed or hierarchical architecture. Centralized implementation is the simplest because a central processing node fuses the data collected by the other sensor nodes. Thus, any erroneous report can be easily detected. However, this architecture is inflexible to any changes in the network and the workload is concentrated at a single point.

In a distributed or decentralized architecture, data fusion occurs locally at each node on the basis of locally sensed data and the information obtained from the neighbor nodes. This architecture is scalable and can continue to work even when new nodes join or depart the network or any dynamic changes in the network occur.

A hierarchical structure can be formed with the sink as the root of the structure and sensor nodes as the leaf nodes. Sensed data move from the lower levels to the higher levels and are fused while moving. Thus, workload is uniformly distributed among the nodes.

Challenges of Data Aggregation

The major challenge in data aggregation, which is actually also a primary goal of this process, is converting raw data to knowledge. Thus, semantics of the data need to be extracted. The data is spatially distributed and correlated. Moreover, because of noise and other types of transmission errors, there are chances of erroneous data and outliers. Also a particular sensor node is able to sense data in continuous manner, and hence it usually generates a data stream, i.e., a sequence of digitally encoded coherent signals. So aging of data is another issue in a sensor network and extracting knowledge from this rich and massive source of data is really challenging.

Data aggregation schemes, particularly those used for energy conservation, will be discussed in Chapter 4.

3.5   Content-based Naming

Names usually refer to "things," i.e., nodes, networks, data or transactions. A name can be unique and globally defined or it can be unique within the local part of the network only. On the other hand, an address provides the information needed to find these things. Addresses can also be uniquely defined in a global scenario or they can be locally defined. Addresses are often hierarchical, because of their intended use in routing protocols, such as IP addresses. Some type of mapping scheme is required to map between names and addresses.

Traditional practice in fixed, wireless infrastructured and in wireless ad hoc networks is to denote individual nodes by their identity. When addresses are not given a priori, they are determined "in the field" using some addressing schemes. Various addressing schemes, global, network-wide or local, have been proposed and many of them are used by the communication protocols like 802.11, 802.15.4 and 6LoWPAN.

In wireless sensor network, the popular idea is to use *content-based addresses* to enable data-centric routing. In traditional networks, the general query is "Get information from Node <ID>," i.e., in traditonal networks, the application seeks information from a particular device. On the other hand, in a sensor network, the information is important, and the device which is sending the information is not. For example, a query can be "whether a four-legged, yellow animal was observed in region A." A number of sensor devices can be deployed in region A. It is immaterial which sensor device picks up the

TABLE 3.1

Common operators and their meanings

Operator Name	Meaning
EQ	Matches if sensed value is equal to given value
NE	Matches if sensed value is not equal to given value
LT	Matches if sensed value is less than the given value
GT	Matches if sensed value is greater than the given value
LE	Matches if sensed value is less than or equal to given value
GE	Matches if sensed value is greater than or equal to given value
EQ_ANY	Matches anything, *value* is meaningless
IS	Specifies a literal attribute

information and reports back to the application via the sink node. Only, the information is important.

In a sensor network, nodes have limited lifetimes, and so few nodes in a region may die quickly or may be destroyed for some reason. However, information must still be gathered. If nearby nodes are available, they can send sensed data. Thus, instead of describing involved nodes, the content about which the query has been forwarded is referred. The classical option is to define a naming scheme on top of IP addresses, which is pursued by middleware systems now.

An attribute-based naming scheme has been proposed [12]. Attribute-based schemes are generally built over an underlying network-based addressing scheme. In sensor networks, low-level communication can be made using attributes which are not part of the network topology and defined within the application only. The scheme exploits knowledge of sensor data types. Attributes can be the location of a node, types of the sensor nodes, or a range of values in a certain type of sensed data.

The scheme proposed in [12] has been implemented along with *Directed Diffusion* algorithm, described in Section 3.2.1.2. In case of *Directed Diffusion*, interest is propagated in the form of metadata. Thus, each propagated interest describes relevant data or event that the interested node (probably the sink) must obtain. Nodes match these interests with their locally observed data. Therefore, each interest is implemented in terms of attribute-value pair. The format suggested in [12] is: $< attribute - value - operation >$. For example, an application interested in finding temperatures greater than $30^{o}C$ can express its interest as $< TEMP, 30^{o}C, GT >$. Attributes can be any sensed events, such as temperature, blood pressure, humidity etc. A table of operators along with their meanings has been suggested in [12].

Another option in sensor network is *geographic addressing*. Addresses can be expressed by denoting physical positions of the nodes, e.g., coordinates of the event. Thus, an attribute is expressed as a single point in terms of (x, y)

cordinates (in three-dimensions including z). In some applications, geographic location can be expressed as a circle or sphere centered around a given point, or a rectangle defined by two corner points, or a polygon defined by a list of points. Geographical addressing is a special case of content-based addressing.

Content-based addressing is important in sensor networks, as it eliminates the overhead of communication required for resolving name bindings. This also enables activation of application-specific processing inside the network, allowing data reduction near where data is generated.

3.6   Summary

Information gathering is the major task in wireless sensor networks. Sensor devices sense application parameters from the environment and forward these data towards base stations or sinks. Sensor devices form an ad hoc network and therefore reactive protocols are more suitable for data forwarding. Wireless sensor network applications implement data-centric routing rather than id-centric routing used by traditinal wireless network routing algorithms. Based on the application requirements, data forwarding algorithms can be either flat-based or hierarchical. Some important flat-based and hierarchical routing algorithms are discussed in this chapter.

Many applications require forwarding of data towards a specific location (a node or a region); thus, localization of sensor nodes is another important issue. Usually, there are two steps implemented by most localization algorithms: distance estimation and distance combining. Most localization schemes depend on beacons or landmarks whose locations are generally known. Beacon-free algorithms are also discussed in this chapter. Geographical routing protocols are discussed in the later part of this chapter.

Data aggregation and fusion are other requirements in wireless sensor networks. Data aggregation techniques combine the data coming from different sources to eliminate redundancies and minimize number of transmissions towards the sink. This chapter gives an overview of data aggregation. It is discussed in detail in the next chapter. Finally, an overview of content-based naming is given which is another requirement for in-network processing and data-centric routing.

Bibliography

[1] J. N. Al-Karaki and A. E. Kamal. Routing techniques in wireless sensor networks: a survey. *Wireless communications, IEEE*, 11(6):6–28, 2004.

[2] Brad B. Karp and H. T. Kung. Gpsr: Greedy perimeter stateless routing for wireless networks. In *Proceedings of the 6th annual international conference on Mobile computing and networking (MobiCom'00)*, pages 243–254. ACM, 2000.

[3] J. Bachrach and C. Taylor. Localization in sensor networks, 2005.

[4] D. Braginsky and D. Estrin. Rumor routing algorthim for sensor networks. In *Proceedings of the 1st ACM International Workshop on Wireless Sensor Networks and Applications*, WSNA '02, pages 22–31, New York, NY, USA, 2002. ACM.

[5] J. Bruck, J. Gao, and A. Jiang. Map: Medial axis based geometric routing in sensor networks. In *Proceedings of the 11th Annual International Conference on Mobile Computing and Networking*, MobiCom '05, pages 88–102, New York, NY, USA, 2005. ACM.

[6] M. Caesar, M. Castro, E. B. Nightingale, G. O'Shea, and A. Rowstron. Virtual ring routing: Network routing inspired by dhts. *SIGCOMM Comput. Commun. Rev.*, 36(4):351–362, August 2006.

[7] L. Doherty, K.S.J. pister, and L. El Ghaoui. Convex position estimation in wireless sensor networks. In *INFOCOM 2001. 20th Annual Joint Conference of the IEEE Computer and Communications Societies. Proceedings.* Volume 3, pages 1655–1663. IEEE, 2001.

[8] Q. Fang, J. Gao, L. J. Guibas, V. de Silva, and L. Zhang. Glider: Gradient landmark-based distributed routing for sensor networks. In *INFOCOM 2005. 24th Annual Joint Conference of the IEEE Computer and Communications Societies. Proceedings.* Volume 1, pages 339–350. IEEE, 2005.

[9] R. Flury, S. V. Pemmaraju, and R. Wattenhofer. Greedy routing with bounded stretch. In *INFOCOM 2009, IEEE*, pages 1737–1745, April 2009.

[10] W. B. Heinzelman, A. P. Chandrakasan, and H. Balakrishnan. An application-specific protocol architecture for wireless microsensor networks. *Wireless Communications, IEEE Transactions on*, 1(4):660–670, 2002.

[11] W. R. Heinzelman, A. Chandrakasan, and H. Balakrishnan. Energy-efficient communication protocol for wireless microsensor networks. In *System Sciences, Proceedings of 33rd Annual Hawaii International Conference*, pages 10–pp. IEEE, 2000.

[12] C. Intanagonwiwat, R. Govindan, and D. Estrin. Directed diffusion: A scalable and robust communication paradigm for sensor networks. In *Proceedings of 6th Annual International Conference on Mobile Computing and Networking*, MobiCom '00, pages 56–67, New York, 2000. ACM.

[13] F. Kuhn, R. Wattenhofer, and A. Zollinger. Asymptotically optimal geometric mobile ad hoc routing. In *Proceedings of 6th International Workshop on Discrete Algorithms and Methods for Mobile Computing and Communications*, pages 24–33. ACM, 2002.

[14] F. Kuhn, R. Wattenhofer, and A. Zollinger. Worst-case optimal and average-case efficient geometric adhoc routing. In *Proceedings of 4th ACM International Symposium on Mobile Ad Hoc Networking and Computing*, pages 267–278. ACM, 2003.

[15] J. Kulik, W. Heinzelman, and H. Balakrishnan. Negotiation-based protocols for disseminating information in wireless sensor networks. *Wireless Networks*, 8(2/3):169–185, 2002.

[16] A. Manjeshwar and D. P. Agrawal. TEEN: a routing protocol for enhanced efficiency in wireless sensor networks. In *International Parallel and Distributed Processing Symposium*, volume 3, pages 30189a–30189a. IEEE Computer Society, 2001.

[17] A. Manjeshwar and D. P. Agrawal. APTEEN: A hybrid protocol for efficient routing and comprehensive information retrieval in wireless sensor networks. In *Parallel and Distributed Processing Symposium, International.* Volume 2, pages 0195b–0195b. IEEE Computer Society, 2002.

[18] D. Moore, J. Leonard, D. Rus, and S. Teller. Robust distributed network localization with noisy range measurements. In *Proceedings of 2nd International Conference on Embedded Networked Sensor Systems*, SenSys '04, pages 50–61, New York, 2004. ACM.

[19] R. Nagpal, H. Shrobe, and J. Bachrach. Organizing a global coordinate system from local information on an ad hoc sensor network. In *Information Processing in Sensor Networks*, pages 333–348. Springer, 2003.

[20] D. Niculescu and B. Nath. Ad hoc positioning system (APS). In *Global Telecommunications Conference, 2001. GLOBECOM'01. IEEE*, Volume 5, pages 2926–2931. IEEE, 2001.

[21] D. Niculescu and B. Nath. Ad hoc positioning system (aps) using aoa. In *INFOCOM 2003. 22nd Annual Joint Conference of IEEE Computer and Communications.* Volume 3, pages 1734–1743. IEEE, 2003.

[22] A. Rao, S. Ratnasamy, C. Papadimitriou, S. Shenker, and I. Stoica. Geographic routing without location information. In *Proceedings of 9th Annual International Conference on Mobile Computing and Networking*, pages 96–108. ACM, 2003.

[23] Z. Rehena, S. Roy, and N. Mukherjee. A modified spin for wireless sensor networks. In *Communication Systems and Networks (COMSNETS), 2011 Third International Conference*, pages 1–4, Jan 2011.

[24] A. Savvides, C. C. Han, and M. B. Strivastava. Dynamic fine-grained localization in ad hoc networks of sensors. In *Proceedings of 7th Annual International Conference on Mobile Computing and Networking*, Mobi-Com '01, pages 166–179, New York, 2001. ACM.

[25] Y. Shang, W. Ruml, Y. Zhang, and M. P. Fromherz. Localization from mere connectivity. In *Proceedings of 4th ACM International Symposium on Mobile Ad Hoc networking and Computing*, pages 201–212. ACM, 2003.

[26] Y. Yu, R. Govindan, and D. Estrin. Geographical and energy aware routing: A recursive data dissemination protocol for wireless sensor networks. Technical report, 2001.

[27] Z. Zhong and T. He. MSP: multi-sequence positioning of wireless sensor nodes. In *Proceedings of 5th International Conference on Embedded Networked Sensor Systems*, pages 15–28. ACM, 2007.

4

Energy Management in WSN

4.1 Introduction

A sensor node is a tiny device with basic components like a sensing subsystem for gathering data from the surrounding environment, a processing subsystem for low level local computation and storage, a communication subsystem for wireless transmission and a power subsystem containing the source of energy. However, the power source is generally a battery with limited energy and little provision for recharging. Still it is expected that the sensor network with such resource-constrained nodes will be able to meet the application requirements within its lifetime. Most of the applications require month-long or year-long execution, thereby implying that the wireless sensor network (WSN) should have such a long lifetime. Hence the important requirement is to prolong the network lifetime. But it is easier said than done.

Now let us look into the various ways a sensor node spends its energy. Here we consider the sensor network model depicted in Figure 1.2 in Chapter 1. The WSN consists of one sink node (or base station) and large number of sensor nodes deployed over a geographic region considered to be the sensing field. Sensor nodes sense data from the sensing field and transfer these to the sink through a multi-hop communication paradigm. Experimental results show that energy consumption during data transmission by a node is the most expensive task. Local data computation or data processing consumes significantly less energy, the order being one thousandth of that required for transmitting single bit of data or information [34] [37]. The sensing subsystem consists of different types of sensors, depending upon the application. Though sensing subsystems do not usually consume much energy, some particular types may end up requiring more energy. Hence energy consumption by the sensing subsystems may be considered to be negligible barring a few applications. Existing energy-saving techniques mainly focus on the operations of a sensor node that include local computations, designing of the routing protocols and sensing subsystems for frequency of gathering samples and energy-reducing sampling techniques.

The architecture of a sensor node is shown in Figure 1.2 in Chapter 1. The main components of the node are: (i) sensing subsystem (sensor) for data acquisition, (ii) processing subsystem that consists of a microcontroller (microcontroller unit) and memory (external memory) for local computation,

TABLE 4.1
Energy consumption of Mica2 mote

Operation	nAh
Transmitting a packet	20.000
Receiving a packet	8.000
Radio listening for 1 ms	1.250
Operating sensor for one sample (analog)	1.080
Operating sensor for one sample (digital)	0.347
Reading a sample from ADC	0.011
Flash read data	1.111

(iii) radio subsystem (transceiver) for wireless data transmission and reception and (iv) power subsystem (supply voltage). As already mentioned, the communication subsystem (transceiver) consumes much more energy than the computation subsystem. Power consumption of the sensing subsystem depends upon the type of the application. The radio subsystem consumes power more or less in the same order for transmission, reception and idle states. But power consumption in the sleep state drops at least one order in magnitude. Based on this nature of power consumption and the architecture of the node as mentioned in Chapter 1, approaches for reducing power consumption are exploited.

Table 4.1 shows the consumption of power for different operations by the Mica2 mote. Energy consumption of the TelosB mote is mentioned in Chapter 6.

The main energy-enabling techniques as identified by Anastasi et al. in [5] are: (i) duty cycling, (ii) data-driven approaches and (iii) mobility-based approaches.

One of the effective methods of energy conservation is to put the radio transceiver in sleep mode whenever communication is not required and then resume use as required. This behaviour is referred to as duty cycling. Thus, duty cycle is referred to as the time interval during which sensor nodes are active in their lifetime. So, any duty cycling scheme will be supplemented with a sleep and wakeup type of scheduling algorithm. The basic pattern of this *sleep* and *listen* or *wakeup* is shown in Figure 4.1.

FIGURE 4.1
Sleep and wakeup

Data-driven approaches take care of data sampling rate and subsequent filtering and/or in-network processing activities along with energy consumption of the sensing subsystems. Mobility-based schemes can be classified as mobile-sink and mobile-relay based schemes. Since static wireless sensor networks are considered, mobility-based schemes are not discussed in this chapter.

4.2 Duty Cycling

Energy conservation using the concept of duty cycling means reducing or optimizing the use of a radio transceiver to lengthen the lifetime of a node. One of the ways of achieving this is by selecting nodes in turn to execute the activities. In sensor networks, this is feasible since the network may generally consist of a large number of nodes and hence connectivity will not be a problem. Depending on deployment policy and application requirement, locations of sensor nodes may be known and utilized. Otherwise, nodes may be activated dynamically such that data is collected from the entire sensing area and connectivity is ensured for data to make its way to the sink. This category of duty cycling may be dependent on or independent of routing protocols. The other method that selects nodes dynamically is the efficient use of sleep and wakeup patterns of the nodes in different algorithms. This category basically utilizes power management techniques.

4.2.1 Independent Strategies

Energy-conserving strategies that can be coupled with any routing protocol, that is, strategies that are independent of routing protocols are topics in research work. A few such strategies are briefly mentioned below.

4.2.1.1 Geographical Adaptive Fidelity

In GAF (geographical adaptive fidelity) [53], the sensing area is divided into virtual grids. For any two adjacent grids X and Y, all nodes in X are able to communicate with nodes in Y, and vice versa (shown in Figure 4.2).

Only one node in a virtual grid needs to be active at any point of time. Thus nodes have to coordinate the periods of sleeping or periods of remaining active among themselves. A node would be in the discovery (exchanges discovery message) or active state (periodically rebroadcasts discovery message) and can change its state to sleep when it finds another node (in the same grid) is active. Nodes in the sleeping state wake up after some interval and go back to the discovery state. GAF may not affect the routing protocol with respect to packet loss and message latency, but the entire sensing area may not be

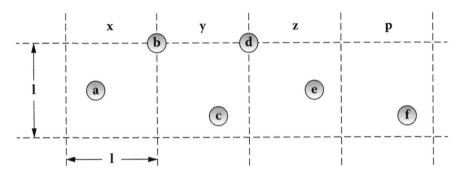

FIGURE 4.2
GAF showing virtual grids

covered. The other restriction imposed is that all nodes in a virtual grid must be able to reach any node in an adjacent grid.

GAF is also mentioned in Section 4.4.2.1 from the perspective of energy-aware location-based routing protocol.

4.2.1.2 Geographic Random Forwarding

The geographic random forwarding (GeRaF) protocol is described in [60], [59]. A node first broadcasts a packet of data containing its own location and the location of the intended receiver, after which receiver-initiated forwarding phase starts.

Figure 4.3 shows the working of the scheme, where S is the source and D is the destination. Nodes in the region with higher priority contend for for-

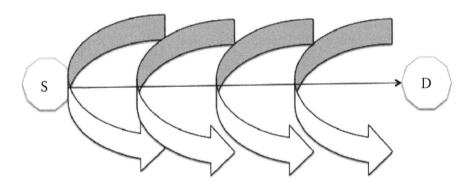

FIGURE 4.3
Geographic random forwarding

warding. Eventually the packet reaches its intended destination. Thus GeRaF requires position information only.

4.2.1.3 Adaptive Self-configuring sEnsor Network Topologies

In adaptive self-configuring sensor nEtworks topologies (ASCENT) [8] some nodes are active initially while others are passive. Passive nodes only collect information and do not cooperate with others. Active nodes forward data and control information. However if high message loss from a source is experienced by a sink, it requests passive nodes to become active. An active node may also solicit help from passive nodes in such situations. However, energy conservation may not increase proportionally with node density.

4.2.2 Dependent Strategies

Energy-conserving strategies that have to be coupled with specific routing protocols, that is, strategies that are dependent on routing protocols (for different information etc.), are also subjects of research work. One such strategy is briefly mentioned below.

4.2.2.1 Span

Span [9] elects coordinators of all nodes in the network. These nodes stay awake and perform routing, whereas other nodes periodically check whether they need to become coordinators. A coordinator eligibility rule is followed by span.

This election algorithm requires neighbour and connectivity information so that a node may decide whether it should become a coordinator or not. This is where the dependence on the routing protocol lies. It is desired that nodes with higher expected lifetimes shall become coordinators and the number of coordinators should be optimal. If multiple nodes try to become coordinators at the same time, the coordinator announcement is deferred by a random backoff delay.

Both the above categories of protocols can typically increase the network lifetime by a factor of 2 to 3 with respect to a network where nodes are always active [13]. But these protocols must be coupled with other energy conservation techniques described in the next section.

4.2.3 Independent Sleep/Wakeup Schemes

The schemes described in this subsection are implemented as independent protocols on top of MAC layer protocols.

Typically, the best practice should be that a node will wake up just in time for receiving and otherwise will remain asleep. This approach will consume the least energy and hence is suitable for wireless sensor network nodes. This type of scheme is classified as on-demand.

4.2.3.1   Sparse Topology and Energy Management

In sparse topology and energy management (STEM) [43], two different radio channels (one for wakeup and the other for data transmission) are maintained.

Each node periodically turns on its wakeup radio for some time at regular intervals. When a source node (initiator) wants to communicate with some other node (target), it sends a stream of periodic beacons on the wakeup channel. When the target receives the beacon, it sends a wakeup acknowledgement and also turns on its data radio channel. However, in case of collision on the wakeup channel, the corresponding node turns on its data radio without sending an acknowledgement. The initiator repeats the beacon transmission in the wakeup until and unless wakeup acknowledgement from the target is received. This protocol is also referred to as STEM–B. The other variation of this protocol is STEM–T [42] which uses a tone that wakes up the entire neighbourhood.

STEM may be used in combination with GAF described in Section 4.2.1. It is found to reduce energy consumption to about 1% or an increase in network lifetime by a factor 100.

4.2.3.2   Pipelined Tone Wakeup

Pipelined tone wakeup (PTW) [55] also uses a wakeup tone to wake up all nodes in neighbourhood of the sender. The duration of the tone should be long enough such the nodes are able to hear the tone when they wakeup at regular intervals from their sleep. Figure 4.4 shows the pipelined wake up procedure. After sending the tone, the sender also sends a notification on the data channel to notify the intended receiver. The receiver sends an acknowledgement on the wakeup channel and the source starts sending data. Thus other neighbour nodes of the sender can go to sleep again. Simultaneously, the receiver (assuming it to be an intermediate hop for the data) now will wake up its neighbour to forward the data packets.

The wakeup radio will also consume some energy and in order to reduce use, another scheme is proposed.

4.2.3.3   Radio-triggered Power Management Scheme

The basic concept of this scheme [17] is to trigger the activation of the sensor node by the energy contained in the wakeup signal or tone. A radio-triggered circuitry is used to capture the energy and use it to trigger an interrupt for waking up the node. One of the drawbacks of this approach is the limitation on the maximum distance from which a wakeup signal or tone can be sent.

Schemes proposed in the literature rely on synchronous senders and receivers. Hence these schemes require that both sender and receiver are in the wakeup mode at the same time. These are categorized as scheduled rendezvous schemes. The protocols differ in the way the nodes wake up and sleep during their lifetimes.

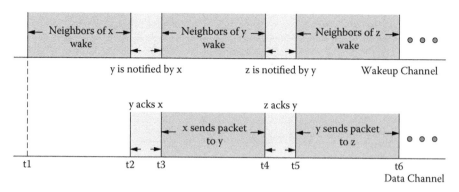

FIGURE 4.4
Pipelined tone wakeup

4.2.3.4 Fully Synchronized Pattern

In this scheme [22] the nodes wake up at regular intervals and remain active for another fixed time interval. Then, they again go to sleep until the next wakeup instant. This scheme is used in several practical implementations including TinyDB [27] and TASK [7]. However, all nodes will try to transmit (after sleeping for a period) and this may cause a collision. This scheme applies to both flat and structured sensor networks. Thus it may be thought to work for a routing tree or data gathering tree organization as well.

4.2.3.5 Staggered Wakeup Pattern

The staggered wakeup pattern [22] assumes nodes form a data gathering tree. Hence nodes located at different levels of the tree should wake up at different times with partial overlapping between parents and children (that is, nodes belonging to adjacent levels). The active portions of different levels are arranged in such a way that the active period a node uses to receive packets from its children is adjacent to the the active period it uses to send packets to its parent in Figure 4.5. Here collisions will be far fewer and the active period may also be shortened. This scheme is suitable for data aggregation also.

Several other scheduled rendezvous schemes are also discussed in literature. These include adaptive and low latency schemes [6] and flexible power scheduling (FPS) [20], derived from on-demand TDMA to name a few.

4.2.4 Asynchronous Schemes

There are a number of other schemes proposed in literature that allow sensor nodes to wake up according to their own schedules independent of oth-

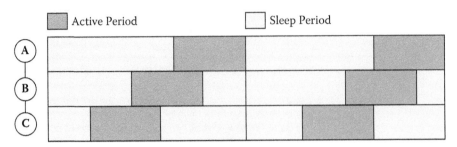

FIGURE 4.5 (SEE COLOR INSERT)
Staggered wakeup

ers. These schemes are asynchronous schemes. The concept is based on IEEE 802.11 Power Saving Mode (PSM) [46] for single-hop ad hoc networks.

4.2.4.1   Asynchronous Wakeup Protocol

The designers of the asynchronous wakeup protocol (AWP) [44] generate asynchronous wakeup mechanisms by viewing the protocol as a combinatorial design problem. This is because neighbour nodes must have some overlap in their active periods for communication. For a given number of neighbours, all nodes have the same duty cycle that indicates wakeup repetition, number of active slots, and number of slots for which any two schedules overlap.

4.2.4.2   Random Asynchronous Wakeup

The random asynchronous wakeup (RAW) protocol [32] is a combination of a geographic routing protocol and a random wakeup scheme. RAW considers the fact that there may be more than one possible forwarder (since there are generally multiple routes in dense WSNs) among the neighbour nodes. Hence the packet is sent to any of the active neighbors in the forwarding candidate set, that meets a pre-specified criterion. Whenever a node wakes up, it does an active neighbour discovery for possible transmission. Thus, RAW may not guarantee that an active neighbour would be found during every wakeup interval and hence packet forwarding may not take place then. However, AWP guarantees packet delivery in every wakeup schedule.

Protocols like STEM–B and PTW discussed earlier can also be categorized as asynchronous schemes.

In STEM–B, the sender transmits a stream of periodic beacons (discovery message) and the receiver's listening time should be at least as long as the time to send a discovery message and the interval between two discovery messages.

Similarly, in PTW, the sender transmits a single long discovery message (tone). Here the discovery message has to be at least as long as the listening time of the receiver.

Generally asynchronous protocols are easier to implement and do not need tight synchronization among nodes. But the nodes may have to wake up more frequently than they would in other schemes.

4.2.5 TDMA-based MAC Protocols

There are several MAC layer protocols that are energy-conscious and focus on power management issues.

As is well known one of the channel access methodologies is time division multiple access (TDMA), in which time is divided into periodic frames, each frame consisting of a few slots. Channel access by nodes is performed on a slot-by-slot basis. Hence a node needs to turn its radio on only during its individual slot. This activity essentially reduces energy consumption. These schemes are categorized as TDMA-based MAC protocols.

4.2.5.1 Traffic-adaptive Medium Access Protocol

In traffic-adaptive medium access protocol (TRAMA) [40], time is divided into random-access period (*signalling slot*) and scheduled-access period (*transmission slot*). Nodes using TRAMA exchange their two-hop information and transmission schedule in particular. Then nodes are selected (chronologically) for transmission and receiving. TRAMA uses the neighbourhood protocol (NP), schedule exchange protocol (SEP) and adaptivity election algorithm (AEA). Traffic-based scheduling is used so that slots are not wasted when a node does not have data to send. A node that is not a receiver switches to low-power standby radio mode.

During the *signaling slot* one-hop neighbour information is propagated by NP for obtaining consistent two-hop topology information. Channel acquisition by nodes is contention-based during this slot, whereas collision-free data exchange and schedule propagation are done in the *transmission slot*. A node announces its schedule using SEP before starting actual transmission. Apart from maintaining consistent schedule information, SEP also updates schedules periodically.

AEA obtains information from NP and SEP and thereby selects transmitters and receivers for a particular time slot that achieve collision-free transmission. This is required to achieve energy efficiency in the collision-free transmission schedule. Otherwise with transmitters and receivers belonging to different slots, energy efficiency may not be achieved and rather may result in waste of energy.

Simulation results for TRAMA show that the protocol achieves significant energy savings and higher throughput when compared with some other contention-based protocols.

4.2.5.2   Flow-aware Medium Access

Flow-aware medium access (FLAMA) [39] is generally used for periodic monitoring applications. It is based on the concept of TRAMA. FLAMA sets up flows and uses pull-based mechanisms. Since the situation is generally stable in periodic reportings, data are transferred only on request. This eliminates the overhead associated with the exchange of possibly unnecessary traffic information. The pull-based approach transfers data only when an explicit request is made.

4.2.6   Contention-based MAC Protocols

MAC protocols basically deal with channel access functionalities which can be coupled with sleep/wakeup algorithms. These are contention-based MAC protocols.

4.2.6.1   Sensor MAC

Sensor MAC (SMAC) [56] follows the scheduled rendezvous scheme. Novel techniques are used for lesser energy consumption and self-organization. SMAC uses an adaptive listening scheme for multi-hop sensor networks. Though all nodes are free to choose their own listen and sleep schedules, synchronization among neighbouring nodes is preferred. Nodes exchange their schedules by broadcasting *SYNCH* packets periodically. Hence a node A will communicate with another node B only when B is listening or awake. However, if multiple nodes want to communicate with B at the same time, contention and collision will occur. SMAC resolves this by virtual and physical carrier sensing and the RTS/CTS exchange for a hidden terminal problem. Virtual carrier sensing is done by including a duration field in each transmitted packet that is called a *network allocation vector* (NAV). Physical carrier sensing is done in the usual way by listening to a channel for possible transmission at the physical layer.

Broadcast packets are generally sent with request-to-send (RTS) and clear-to-send messages. The full sequence of RTS, CTS, data and acknowledgement (ACK) is followed for unicast packets between a given pair consisting of a sender and receiver. After successful RTS and CTS, the two nodes use their normal sleep times for data packet transmission. However, they do not follow their sleep schedules until transmission is finished.

SMAC allows periodic neighbour discovery. A node without a neighbour tries to discover neighbours more aggressively than do nodes with neighbors. Some type of time synchronization scheme is also required in SMAC since neighbouring nodes coordinate their sleep schedules.

Adaptive Listening was introduced in SMAC to improve the latency caused by the periodic sleeping in a multi-hop network. The concept is that a node that overhears its neighbours' RTS or CTS (ideally) transmissions, wake-up for a short time when the transmission ends. If the node is the next-hop node, it immediately receives the data. However, the node will go back to sleep again if it does not receive a transmission during this adaptive listening and then wake up at the scheduled listening time.

Figure 4.6 shows the timing relationship between a receiver and senders.

4.2.6.2 Timeout MAC

The novel idea behind timeout MAC or TMAC [50] is the introduction of an *adaptive* duty cycle by dynamically ending the active part of the duty cycle. This reduces the energy spent on idle listening. The usual pattern is periodic wakeup of a node to communicate with its neighbours, and then going to sleep again until the next frame. Collision avoidance and reliable transmission are provided by the usual RTS/CTS/DATA/ACK scheme, which the nodes use for communication among themselves.

A node continues listening for potential transmissions as long as it is in its active period. The active period ends when no activation event occurs for a time *TA*. Activation may be the firing of a periodic frame timer, receiving data, sensing a communication in a channel, ending its own transmission of data or an acknowledgement, or the end of data transmission by a neighbor. Through observation, the authors arrived at a lower limit on *TA* consisting of a combination of contention interval, length of RTS package and time interval between the end of the RTS packet and the beginning of the CTS packet.

An interesting find during the implementation of TMAC protocol is that during high load in the system, nodes communicate without sleeping, but

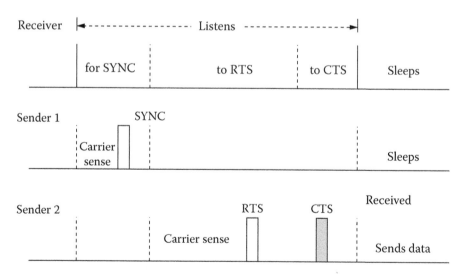

FIGURE 4.6
Timing relationship between senders and receiver [56]

during low load, nodes wisely use their radios for as little as 2.5% of the time, and may save up to 96% of the energy compared to a traditional protocol.

4.2.6.3   Berkeley MAC

Berkeley MAC (BMAC) [33] is a low complexity and low power MAC protocol which is quite popular and comes bundled with the TinyOS operating system [18]. Any MAC protocol must have the ability to accurately determine whether the channel is clear for transmission, referred to as clear channel assessment (CCA). BMAC uses software automatic gain control and signal strengths are sampled when the channel is assumed to be clear to get an estimate of the noise floor. Basic channel access control features like backoff scheme, an accurate channel estimation facility and optional acknowledgements are implemented in BMAC.

An asynchronous sleep/wake scheme based on periodic listening called low power listening (LPL) is also implemented to have a low duty cycle. Nodes periodically wake up to check the channel for activity. The wakeup time is fixed but the intervals after which the nodes wake up may be specified by the application. BMAC packets consist of long preambles and payloads. When a node detects that a channel is active, it also stays active and receives both the preamble and the payload. Thus nodes need not be synchronized.

A brief critical analysis of SMAC and BMAC is given here.

- SMAC is designed using the classical approach of RTS and CTS mechanisms for channel arbitration and hidden terminal avoidance. This also enables synchronization with neighbours and message fragmentation for efficiently transferring bulk data. Synchronization with neighbours helps in low power operation. SMAC functions as a protocol for accessing medium and also works as network and organization protocol.

- BMAC has limited media access functionality that uses clear channel assessment (CCA) and packet backoffs for channel arbitration, link layer acknowledgements for reliability and low power listening (LPL) for low power communication. However, it is only a link layer protocol, and unlike SMAC, does not have the functionality of network services like organization, synchronization, and routing. For services like message fragmentation and hidden terminal support, BMAC has an additional set of interfaces that allow services to tune to its operations including CCA, acknowledgements, backoffs, and LPL. Protocols can be built on BMAC to locally decide on optimizing power consumption, throughput, latency, fairness or reliability.

Although energy efficient, the MAC protocols suffer from sleep latency and hence are not suitable for multi-hop networks.

4.2.6.4   Energy-efficient Low-latency DMAC

DMAC protocol [26] is designed to be an energy-efficient and low-latency MAC optimized for data gathering trees in wireless sensor networks. DMAC uses a staggered active and sleep schedule that enables continuous data forwarding on multi-hop path. This is done by staggering the activity schedule of nodes on the multi-hop path such that the wakeup is sequentially like a chain reaction.

The sending and receiving periods are of same length and include receiving and sending of explicit acknowledgement messages, whereas the sleep period depends on the depth of the node in the data gathering tree. Collisions during transmission periods of nodes at the same level are also taken care of. Local synchronization is needed since a node has to be aware of its neighbour's schedule. Aggregated rate at a node may be larger than what it can handle so data prediction is employed.

The interference between nodes with different parents is handled by the use of an explicit control packet called *more-to-send* (MTS). The MTS packet is very short and contains destination, local identification and a flag. When the flag is set to 1, MTS is the *request* MTS. The MTS with flag set to 0 is called *clear* MTS. *Request* MTS is sent when a node is unable to send or it receives the same item from its children. *Clear* MTS is sent when a node's buffer is empty and the node earlier sent a *request* MTS to its parent did not send a *clear* MTS for that, and all *request* MTS packets from its children were handled.

Energy consumption increases due to MTS and longer slots, but the protocol achieves low latency in WSNs having multiple sources and there are minimal additional energy costs.

4.2.6.5   IEEE 802.15.4

This [19] is a standard for personal area networks (PANs) that is low rate and requires low power. It supports both beacon-enabled mode and non-beacon-enabled mode channel access methods. The beacon-enabled mode uses a superframe structure bounded by beacons, containing an active period and an inactive period. The active period is subdivided into a contention access period (CAP) and collision-free period (CFP). During the CAP, a slotted CSMA/CA algorithm is used whereas in CFP a number of guaranteed time slots (GTSs) can be assigned to each node. During the inactive period, nodes switch to low power state, thereby conserving energy. In the non-beacon-enabled mode, nodes are always in active state and use an unslotted CSMA/CA algorithm.

4.2.7   Hybrid MAC Protocols

The two basic channel access technologies used in MAC protocols are TDMA and CSMA. The challenging concept is to switch the protocol behaviour between TDMA and CSMA, depending upon the level of contention of the medium. This results in hybrid MAC protocols.

4.2.7.1   Probabilistic TDMA

Probabilistic TDMA or PTDMA [12] was proposed for one-hop WLAN scenarios. Time slots are used, and nodes are distinguished as owners and non-owners of a particular time slot. The protocol tries to find out the number of senders in a particular time slot and then access probability of owners and non-owners are determined. Accordingly channel access is adjusted.

The adaptability by PTDMA basically results in MAC working as either TDMA or CSMA, depending on the level of contention in the network. Since PTDMA was initially conceived for one-hop scenarios, it does not consider other network issues like change in topology, errors due to synchronization or interference, which occur commonly in wireless sensor networks.

4.2.7.2   ZMAC

ZMAC [41] is designed as a hybrid scheme combining only the strengths of TDMA and CSMA. It takes CSMA to be the basic MAC scheme, at the same time using a TDMA schedule that tries to enhance contention resolution. The *setup* phase (that runs only once unless there is significant network topology change) performs the following operations sequentially: neighbor discovery, slot assignment, local frame exchange and global time synchronization. The initial costs for the *setup* phase are offset by improvement in throughput and energy efficiency.

Through neighbour discovery protocol, a node is aware of its one-hop and two-hop neighbourhood. Each node in the network maintains its own local time frame that fits its local neighbourhood size, but employs mechanisms that avoid conflict with contending neighbours. Every node also synchronizes to slot 0.

In ZMAC, a node may be either in low contention level (LCL) or high contention level (HCL). If a node receives an explicit contention notification (ECN) message from a two-hop neighbor within a pre-set time interval, it is in HCL. Also a node can send ECN if it experiences high contention. Each node is statically assigned a time slot (TDMA), but unlike TDMA, it can transmit in both its own time slot and slots assigned to other nodes. Likewise, an LCL node can compete to transmit in any slot, but only "owner" nodes in HCL and their one-hop neighbours are allowed to compete for a current slot for channel access.

However owners always have higher priority than non-owners. But if a slot does not have an owner or its owner does not have data to send, non-owners can utilize that slot. Hence, under low contention, non-owners can "steal" time slots from owners. This effectively has the effect of switching between CSMA and TDMA, depending on contention. Thus, performance of CSMA is achieved under low contention, and that of TDMA under high contention. ZMAC is implemented in TinyOS.

TDMA-based MAC protocols are inherently energy-efficient. But these have certain drawbacks too, like limited flexibility and scalability. Hence they are not very popular in real-life sensor networks. Contention-based MAC protocols are robust and scalable, but their energy consumption is higher due to contention and collision. Though hybrid MAC protocols combine the strengths of both, but they are too complex to be implemented in a large network.

4.3 Data-driven Approaches

Techniques using data-driven approaches aim either at reducing data-related activities at nodes or at efficient data acquisition methods to conserve energy.

The data reduction activities may range from performing some computations on data (before being sent to next hop or sink) or schemes like *compression* performed by the sensing nodes or abstract data models that are able to predict data values. Computations on data mainly perform *data aggregation*, which may be averaging of data values, finding maximum and minimum among data values. Basically *data aggregation* is always application-specific. The authors in [18] have done a comprehensive survey of different in-network data processing techniques. *Data compression* techniques are general and there are many well–known techniques. Detailed discussion about these may be found in [45]. *Data prediction* techniques use models that describe the sensing activity. The models are then used to answer queries. The models will reside both at nodes and sink. The features of the specific *data prediction* technique depend on the way the model is built.

For data acquisition to be less energy-consuming, techniques of data sampling must be efficient so that they can cope with the power requirements of sensors, transducers, and other devices.

Approaches relating to *data prediction* and *energy-efficient data acquisition* are briefly described next.

4.3.1 Data Prediction

The models used in *data prediction* reside in the sink and the source nodes. The model at the sink can be used for answering queries only if it represents a phenomenon properly and at that instant. The model at the source node is used for data validation purposes, that is, if the sampled data value falls within the predicted range, the model is considered valid and hence data is actually not communicated to the sink from the source. Thus the importance focuses on the way the models are to be built. The authors in [5] identified three approaches for *data prediction*.

4.3.1.1   Stochastic Approaches

The stochastic approaches use random processes to characterize the sensing phenomena. The use of random processes in turn allows probabilistic models to predict sensed values.

- Ken approach: This scheme [10] assumes a number of models, each at the source and at the sink. The probability density function (pdf) of the model with respect to a set of attributes is obtained after a training phase. The source node transmits sample data to the sink when the model becomes outdated. Thus the model at the sink is also updated. The Ken approach can be custom-made for applications dealing with specific sensing phenomena in which temporal or spatial correlations may have to be exploited. Markov processes and greedy approaches are used by the authors in these situations.

- Approach using dynamic probabilistic model (DPM): This approach [21] is used for obtaining a consistent snapshot of data from a model. The internal (hidden) state of sensor database is obtained via DPM. Thus it can indicate whether the operational state of the node (represented by the model) is working correctly or not. The output of the DPM is stored as a set of weighted samples (particles). The queries are transformed to suit this particle-based representation. These queries can also be optimized. A Monte Carlo algorithm is used for performing particle filtering for updates.

4.3.1.2   Time Series Forecasting

A number of techniques use historical values (time series) obtained by periodic samplings to predict future value in the same series. This is known as time series forecasting. The methods generally used for representing time series are *moving average (MA)*, *auto-regressive (AR)*, and *auto-regressive moving average (ARMA)*, to name a few.

- PAQ [49] is based on the low-order AR model. During the learning phase, nodes store the set of sampled (sensed) values in a queue. When the queue is full, the first instance of the model is built and ready to be sent to the sink. Thus communication between nodes and the sink is not the actual sensor reading but consists of parameters of the model (that is, the coefficients of the AR model). Each model considers a pre-specified error bound. The model is considered to be valid for a particular sensed data value, if the predicted value is within that bound. Otherwise, either the sampled data are marked as outliers or, the model is said to be invalid if sensor readings fall outside the bound repetitively. Outliers may be sent to the sink or simply ignored, whereas the model is to be recomputed. PAQ also proposes a distributed clustering scheme for nodes represented by the same model and same error bound. Clustering helps in conserving energy because cluster-heads are now only involved in communication with the sink regarding models.

- In SAF [48], the AR model is refined to include a trend component. SAF can detect both outliers and inconsistent data. The stability of the model may be improved by the nodes by data filtering to smooth outliers, and enlarging the size of data to decrease impact of outliers. Besides this, nodes may still rebuild models. The centralized clustering scheme used in this work is optimal with respect to the number of clusters.

4.3.1.3 Algorithmic Approaches

There are other approaches that extend *time series forecasting* by an adaptive multi-model selection mechanism. These approaches suggest that nodes keep a set of models with themselves but use only one at a given instant. If the error between sensed data and current model is beyond the allowed threshold, then a new model is chosen and proper update at the sink is also made.

Several other models are also proposed in literature for data prediction. *Algorithmic approaches*, like heuristic or behavioral characterization of sensed phenomena are generally used by these schemes.

- PREMON: In the PREdiction–based MONitoring paradigm [15], the fact that sink may predict sensor reading given the knowledge of sensor readings around its vicinity is utilized. The sensor will transmit its sensed reading to the sink only when it is beyond the pre-specified threshold from the data generated by the prediction model. Another perspective to this work comes from the idea that WSN may be viewed as an optical image with sensor node readings being intensity values of pixels in the image. Our second key observation is that a snapshot of the sensor network may be visualized as an (optical) image in which the readings of individual sensors correspond to intensity values of pixels in the image. Monitoring by sink has the effect of watching a video of sensed values and hence MPEG [1] may be used for compressing this video.

- The buddy protocol [16] establishes a number of buddy groups (like clusters) and the representative (cluster-head) of each group becomes responsible for monitoring and query processing. Each node sends sampled data to its cluster-head by default. Depending on the estimated energy, a node sends a model and the data that is beyond the bound of prediction in PREMON mode. In steady sensing phenomena, this mode is convenient since actual number of data communicated will be reduced and thus energy is conserved.

- Energy–efficient data collection (EEDC): Instead of proposing a model for sink and sensor nodes, a node sets bounds (lower and upper) whose difference represents the accuracy of sensor readings to the actual sensed data. The sink stores such bounds for each sensor node. The nodes send

the sink an update if the sampled data exceeds this accuracy. This is
called source-initiated update. The sink is able to respond from the cache
to user queries when requested accuracy is lower than actual accuracy. If
not, the sink has to request the actual sensed values and updated approx-
imation. This is called consumer-initiated request and update. However,
for efficient consumption of energy, methodologies for selecting optimal
ranges of data have to be developed.

Techniques following *stochastic approaches* perform different operations on
sensed data and hence may have high computational costs, whereas *time series
forecasting* techniques are able to provide satisfactory accuracy with simple
models too. But these techniques require specific types of models suitable to
represent the particular phenomenon. Techniques with *algorithmic approaches*
are always application-specific.

4.3.2 Data Sensing

A large number of emerging applications require power-hungry high-end sensors. A lot of other factors also increase consumption of energy in such applications. Sensing arrays like CCDs or CMOS image sensors or multimedia sensors
generally require high power resources to perform sampling. This is true for
chemical and biological sensors also. Some other sensors like acoustic and seismic transducers require high-rate and high-resolution A/D converters. Some
types of sensors have to send probing signals by using active transducers like
sonar, radar and laser rangers, to acquire data from the sensed phenomenon.
Apart from this, data acquisition time may be so long (a few seconds being
longest) that energy consumption of the sensing subsystem does not remain
negligible.

The requirement therefore, is to reduce energy consumption during the
data sensing or data acquisition. This may be attempted by reducing energy
consumption of the sensing subsystem or by reducing the data sampling frequency which may also lead to minimization of communication.

Data acquisition may be classified according to the sampling techniques
[5] involved.

4.3.2.1 Adaptive Sampling

Adaptive sampling techniques consider data correlation with respect to time.
This means that since rate of change in sampled data is slow, numbers of
acquisitions can be reduced over time. Sometimes sampled data hardly changes
spatially as the investigated phenomenon may not change drastically within
an area covered by sensor nodes. Thus both temporal and spatial correlations
are taken into consideration in this technique.

- Adaptive sampling for snow monitoring: Each WSN application is specific in nature and hence the authors [4] describe an adaptive sampling

approach suitable for forecasting avalanche by snow monitoring. They propose an adaptive algorithm to estimate dynamically the current maximum frequency and thereby use modified CUSUM [31] to set the sampling rate. The algorithm is executed at the sink and the sensor nodes are notified of the sampling rate.

- Backcasting: The authors [52] propose that activation of sensors in the sensing field be done in two phases. In the *preview* phase, only a few nodes are activated to gather a preliminary estimation of the spatial distribution of the sensed phenomenon. This preliminary estimation is also done in successive steps by recursively partitioning the sensing field into subsquares and sending them to the fusion centre, the sink. In the *refinement* phase, the fusion centre backcasts activation messages to cluster-heads of smallest partitions which in turn activate additional nodes in the cluster.

- Correlation-based collaborative MAC protocol (CC-MAC)[51] uses spatial correlation and has two CSMA-CA-based components. An iterative node selection (INS) algorithm is executed by the sink that derives a correlation radius that is broadcast to sensor nodes. A node becomes representative of an area based on this radius. *EMAC* only allows this node to capture the channel and all other nodes within that area stop their transmissions. With no redundant packets *NMAC* gives higher priority to packets by representative nodes than newly generated packets. Nodes other than representative nodes may therefore switch off their sensing subsystems.

- FloodNet [58] is another application-specific (flood warning system) adaptive sampling approach. The *floodNet adaptive routing* (FAR) uses adaptive sampling and energy-aware routing protocol. The routing algorithm uses priority and data importance for selecting forwarder nodes. Priority is based on the residual battery power and energy needed for transmission. Data with high sampling rates are considered to be important since they are associated to critical zones. The nodes that have higher priority and lower data importance are chosen as forwarder nodes. This implies that nodes with higher energy but lesser sampling load are selected. Thus this scheme aims to conserve energy.

- Networked info-mechanical System (NIMS) [38] is another application-specific (environmental monitoring) approach that exploits both temporal and spatial correlation. NIMS is a mobile node with meteorological sensors and an aerial infrastructure that supports the sensor. Both the horizontal and vertical positions of the mobile sensor can be accurately set. First a navigation criterion of the mobile sensor is defined that considers both actuation and sampling costs and characterizes the areas where high variation of the phenomenon is observed. This ensures that places with high error rates are adequately covered. For exploitation of temporal correlations between samples, the system employs an adaptive parameters selection too.

4.3.2.2 Hierarchical Sampling

Hierarchical sampling techniques consider types of sensors (with different resolutions or accuracy, power consumption etc.) each node has and dynamically select the sensor to activate, depending on accuracy and energy consumption.

As an example, we may consider target tracking applications. Initially low-power sensors can be used to detect targets. The detection may not always be very accurate or even correct. Even when detection is correct, the type of the target may not be properly identified. Hence high resolution sensors that are able to capture images or other types of sensors have to be used. These sensors are generally power-hungry. So they may only be utilized when required. This technique is known as *triggered sampling*. In general, energy consumption is optimized.

- Triggered sampling for structure health monitoring and damage detection: The work in [23] describes a *triggered sampling* approach that monitors the health of a structure. The application divides the structure into zones containing different types of sensors. Sensor nodes called m-nodes equipped with accelerometers sample the environment periodically. The other type known as μ–nodes have strain gauges. These sensors are generally sleeping. The m-node also sleeps between activations and cross checks its sensor readings with the readings of its neighbours. The μ-nodes are activated only when some mismatch or other suspicious event arises while checking the readings. The μ-nodes provide fine-grained information and thus eventually report damage.

- Multi-scale sampling uses two main approaches for data accuracy. The concept of activating sleeping high-resolution sensors residing in the area of interest to collect sample data is usually followed by a robotic mobile sensor reaching the location from which more accurate data is required. The work in [47] uses this approach in a fire emergency application scenario. The area of interest has a number of static sensors that monitor the environment. Generally temperature data is sampled and when it is found to exceed a given threshold, nodes send an appropriate message to the sink. The sink, in turn, dispatches a mobile sensor to that location, which collects data from the static sensors and also takes a snapshot. It then returns to the sink and reports its findings.

Model–based active sampling is similar in approach to the concept described in Section 4.3.1. A model of the sensing phenomenon is first built on sampled data that is used to predict future data values. This reduces both the number of data samples and also data to be transmitted to the sink.

- The Barbie Q (BBQ) query system [11] consists of a model and a planner at the sink. The probabilistic model derives a probability density function (pdf) over a set of attributes of a given number of data samples. The attributes vary over time and are modelled as a Markov process. The pdf is able to capture both temporal and spatial correlations and the accuracy of the data sample in terms of probability. The model is continuously updated by combining the pdfs with the data samples so that future values can be predicted correctly. The sink develops the model once the nodes transmit sampled data that are used to obtain the first instances of the pdfs. This initial stored model is updated along with answers to queries and this is the responsibility of the planner. The query plan of the planner has a list of sensors to be queried and the most relevant data to be obtained from them. Since data may be collected via various methods, a planner can choose the method of collection.

  As an example, suppose temperature data of a certain area is to be measured. The planner can directly use temperature sensors to get the data. Another way can be to measure the voltage so that applying correlations (among different attributes) will allow temperature data to be derived. Since voltage measurement is much cheaper than temperature measurement, the planner may as well choose to get the voltage data from some nodes. This reduces the overall power consumption associated with the query. The planner tries to find out the associated observation cost (of the data required in query) by considering both sampling and communication costs. Obtaining an optimal solution is complex, and for this reason the authors propose a polynomial time heuristic effective for practical solutions.

- Adaptive sampling approach to data collection (ASAP): This technique [14] is built upon clustered network where clusters are based on similarity in sensor readings and hop count. Clusters can again be divided into subclusters in which one node (sampler) can gather data and send it to the sink. The samplers are chosen by the cluster-heads by using an initial set of sample data provided by all nodes in the corresponding clusters. The probabilistic models exploiting spatial and temporal correlations are built for each subcluster and sent to the sink. The sink can have actual sensed data from the samplers or predict data by using the model. Both clusters and subclusters are periodically rebuilt. Communication overhead is reduced since model update involves exchange of data within subclusters only.

- Utility-based sensing and communication (USAC): The authors have chosen the application of glacial environment monitoring for describing this protocol in [30]. A limited window linear regression model is used to forecast sample values. In this work sampling frequency is updated at each node following certain conditions. Pre-defined minimum f_{min} and maximum f_{max} values of sampling frequencies are maintained. Whenever the predicted value is beyond the confidence interval, the sampling frequency is increased to f_{max}. Model updates are done when sudden changes are observed in sampled data. The increase in sampling frequency maintains the accuracy of the model. If the predicted value lies within the confidence interval, the sampling frequency is decreased by a factor α with a value between 0 and 1, unless f_{min} is reached. The work also describes a routing protocol that accounts for the energy spent during both sensing and communication. This is done by building an opportunity metric, so that sensors not involved in relaying data can perform the additional task of sampling. Also, routes in which data sampling frequency is lower can be preferred to routes in which data sampling frequency is high, that is, nodes spend more energy for sampling.

With several model-based active sampling approaches, adaptive sampling techniques offer promise, but the techniques considering both temporal and spatial correlation would be more advantageous. These techniques also tend to offer a centralized solution because of the huge complexity of solution which is also a drawback in devices like wireless sensor networks. Distributed approaches considering adaptive sampling techniques are also now being exploited. Hierarchical sampling techniques are application-specific, and hence have limited use. Model-based active sampling techniques hold promise, but they should also be oriented toward computing models in a distributed fashion.

4.4   Energy-aware Routing Protocols

In the earlier Sections 4.2 and 4.3 we discussed a number of enabling techniques in different categories. Many of these techniques can be coupled with any routing protocol. However, there are also a couple of routing techniques that are designed with the aim of optimizing or even minimizing energy consumption of the sensor nodes. This section describes a few such techniques. Techniques already described in Chapter 3 will not be detailed but their relevance to energy optimization is discussed.

4.4.1   Hierarchical Energy-aware Routing

As is well known, a hierarchical structure in a network can be established with clusters that have corresponding cluster-heads with designated tasks. The other non-cluster-head nodes may perform the usual task of sensing. There are mainly two phases for hierarchical routing techniques. Cluster-heads are selected at first, after which routing takes place. The work in [35] presents an analysis and comparison of a number of energy-efficient routing protocols.

4.4.1.1 LEACH

Details of the low energy adaptive clustering hierarchy are covered in Chapter 3. Not all sensor nodes have to send data to the base station or the sink. This saves power consumption since data from sensor nodes are sent only to respective cluster-heads. Thus, cluster-heads consume more energy than other nodes. Hence, the role of a cluster-head is rotated among the nodes in a cluster. The features of selecting cluster-heads and reselecting them at regular intervals saves energy.

Another important feature adopted by a non-cluster-head node is sending data only during its scheduled transmission slot and then turning its radio off (going into *sleep* mode). All these features help conserve energy and increase network lifetime.

4.4.1.2 PEGASIS

PEGASIS, power-efficient gathering in sensor information systems [25], is a near-optimal protocol. The basic idea of the protocol is to produce a chain consisting of nodes closest to each other and form a path to the sink or the base station. The chain is constructed in a greedy fashion. Each of the nodes in the chain will take a turn in communicating the aggregated data to the base station. The node furthest from the base station is first considered for constructing the chain. This is done to ensure that nodes farther from the base station have close neighbours. Since nodes already in the chain cannot be revisited, neighbour distance increases gradually in the greedy algorithm. The chain is reconstructed in the same manner to bypass a dead node.

In each round, each node receives data from one neighbour, fuses that with its own data and transmits to the other neighbour on the chain. A node N will be in some random position p on a chain. Nodes take turns transmitting to the base station and some very basic method is used to determine which node will transmit to the base station in a particular round r. Thus, the leader in each round of communication will be at a random position in the chain. Data fusion is performed at each node except the end nodes in the chain. The length of the fused data packet will be the same as that of the original or received data packet.

Once all nodes in the chain have communicated, a new round will start. This eventually spreads power consumption among all nodes in the network leading to energy conservation and longer lifetimes.

Though PEGASIS does not have any clustering overhead, it has to manage topology dynamically. This is because a sensor node must know the energy status of its neighbours and route its data accordingly. However, PEGASIS scores over LEACH in the way that it eliminates dynamic cluster formation, limits transmissions to the base station, and introduces data aggregation. PEGASIS distributes energy load among the nodes and thus increase the lifetime and quality of the network.

4.4.1.3   TEEN and APTEEN

The TEEN (threshold-sensitive energy-efficient sensor network) protocol is described in Chapter 3. The protocol has certain features in common with LEACH, for example, rotation of cluster-heads. Energy conservation results from clustering. The setting of different threshold levels in TEEN also saves power because the transmission of data depends on threshold levels.

The APTEEN (adaptive periodic threshold-sensitive energy-efficient sensor network) protocol is an improvement over TEEN and is also described in Chapter 3. It provides all the advantages of TEEN and eliminates some of its drawbacks by introducing a regular interval within which nodes send data even when they have nothing worthwhile to send. Both APTEEN and LEACH follow TDMA schedules and hence their radios are turned off when not in use and this also saves energy.

Simulation experiments of TEEN and APTEEN show that both outperform LEACH. TEEN, however, gives best performance among these three. The APTEEN performance level is somewhere between those of LEACH and TEEN with respect to energy dissipation and network lifetime.

4.4.1.4   Hierarchical Power-aware Routing

The wireless sensor network employing hierarchical power-aware routing (HPAR) [24] is assumed to be divided into *zones* that comprise sensor nodes in geographic proximity. Each *zone* decides on the hierarchical routing to be used for its data across other zones. The routing path is chosen such that total remaining power of all nodes in the path is *maximum* over all the *minimum* of remaining power. This path is called the *max–min path*.

The *max–min zPmin* algorithm [24] uses the tradeoff between minimizing the total power consumption (of a path) and maximizing the minimal residual power of the network. The algorithm works as follows. First, the path with the lowest power consumption (Pmin) is found by using the Dijkstra algorithm. Then the path that maximizes the minimal residual power in the network is found. The *max–min zPmin* algorithm tries to optimize the power criteria by setting power consumption to zPmin where $z \geq 1$ to restrict the power consumption for sending single messages to zPmin. Thus at most zPmin power is consumed while maximizing the minimal residual power fraction.

Another algorithm proposed in this work [24], the *zone-based routing*, finds a global path across zones based upon the *max–min zPmin* algorithm. Zone level power estimates are used for routing a message. A global controller manages the zones. *Zone-based routing* is able to drastically reduce the running time to find a route. Also within each zone the local routing path is computed in such a way that it does not decrease the overall power level of the zone.

All the above-mentioned hierarchical routing schemes attempt to minimize the consumption of energy as far as possible. They may have introduced drawbacks like remaining power computation and maintenance of information for comparison but these protocols have emerged to be both only energy-aware and energy-efficient.

4.4.2 Location-based Routing

Location information plays an important role in many WSN applications. Sensor nodes must be made location aware for successful execution of the application. There are a couple of routing schemes based on geographic locations that are power-aware and attempt to conserve energy.

4.4.2.1 Geographic Adaptive Fidelity

In Reference [54], the sensing area is divided into a number of fixed zones, each forming a virtual grid. Figure 4.2 (in Section 4.2.1) is an example of fixed zoning where each zone or cluster is assumed to be equal and square. The selection of the size of the square is dependent on required transmission–power and direction of communication.

The cluster-head receives data from members of its zone or cluster and forwards them to the base station or sink. GAF also tries to keep at least one representative node active during the execution of the application.

A node uses its GPS-indicated location and associates itself with a corresponding point in the virtual grid. However, multiple nodes may be associated with the same point. In that case, they are considered to be equivalent in terms of the cost in routing. This equivalence helps save energy energy since most equivalent nodes in an area may sleep while the non-sleeping node performs the tasks of sensing and communication. Thus, GAF is able to substantially increase network lifetime.

The nodes in GAF determine other neighbours in the grid or participate in routing or sleep. GAF also handles mobility of nodes where a node has to estimate its leaving time send this to its neighbours. The neighbours appropriately wake up before that time expires and at least one of them becomes active for handling the routing activities of the leaving node. GAF is implemented for both non-mobility (GAF basic) and mobility (GAF-mobility adaptation) of nodes.

GAF thus combines the concepts of geographic location-based routing and hierarchical routing.

4.4.2.2 Geographic and Energy-aware Routing

Geographic and energy aware routing (GEAR) [57] is based on directed diffusion algorithm with the restriction of number of interests to be sent to a certain region rather than the entire network. This restriction makes this routing more energy-aware than Directed Diffusion. GEAR uses a heuristic for geographically informed neighbour selection. GEAR was already introduced in Chapter 3. Here we describe the computations and energy involved.

It is assumed that each query packet has a specified target region. Apart from its own location and energy level, a node also knows the information of its neighbours. First the centroid of the target region is computed. A node then tries to route the packet progressively towards the target region, while attempting to balance the energy consumption across all its neighbours.

Each node maintains the *learned cost* value of each of its neighbours towards that region, which it tries to minimize. It will calculate an equivalent *estimated cost* when it does not have the *learned cost*. Whenever at least one neighbour closer to the region is found, it becomes the next-hop node to be picked up. In case there are no such neighbours, the node assumes the existence of a *hole*.

The *learned cost* of the node and its update rule are used to find a way to circumvent the *hole*. Every time a packet is delivered successfully, the *learned cost* value will be the correct one and it will be propagated one-hop away. Propagation of the *learned cost* values through the update rule enables an earlier chance to avoid the *hole*. An energy-aware metric is used in the *estimated cost* function to balance energy consumption that leads to an energy-efficient network.

4.4.3 Data Aggregation-based Routing

Due to large number of sensor deployments in a wireless sensor network, redundant data will be generated often. Transmission of all these data to the base station will generate huge traffic and require unnecessary storage and computation. Data aggregation is one of the popular methods to eliminate this huge redundancy.

Data aggregation combines similar data from different sources according to some particular function depending upon the nature of the data and application requirements. Examples of such functions may be finding maximum, minimum, average and/or some kind of filtering. Thus the technique optimizes the data to be sent to the base station and hence, is energy-efficient. Routing protocols using data aggregation techniques are therefore inherently energy-efficient. Below we discuss a couple of such routing protocols.

4.4.3.1 Virtual Grid Architecture Routing

Virtual grid architecture (VGA) routing [3] utilizes data aggregation and in-network processing. This scheme assumes a fixed topology and builds clusters without the aid of a global positioning system (GPS). The clusters are fixed and non-overlapping with symmetric shapes. Each such cluster has a cluster-head and data aggregation is performed at both local and global levels. Figure 4.7 shows a possible implementation field.

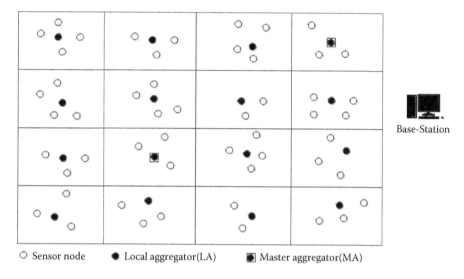

O Sensor node   ● Local aggregator(LA)   ▣ Master aggregator(MA)

FIGURE 4.7
Virtual grid architecture [3]

The cluster-heads are the local aggregators (LAs) and a set of LAs forms global aggregators (GAs). However there has to be an optimal selection of global aggregation points. Various algorithms, like, integer linear programming (ILP), genetics algorithm-based heuristics, k-means heuristic, greedy heuristic, and clustering-based aggregation heuristic (CBAH) [28] are proposed for the selection of MAs from LAs in such a way that it maximizes network lifetime.

Simulation results show that CBAH outperforms LEACH and PEGASIS with respect to network lifetime. At the same time, it also achieves acceptable latency without sacrificing quality.

4.4.3.2 Sensor Aggregates Routing

The authors, in their work [40], describe a set of three algorithms: (i) distributed aggregate management (DAM), (ii) energy-based activity monitoring (EBAM), and (iii) expectation maximization-like activity monitoring (EM-LAM) that construct and maintain sensor aggregates. Generally sensors in a network perform collaborative tasks like target monitoring. A sensor aggregate consists of nodes in a network that satisfy a grouping predicate for a task. The parameters of the predicate depend on the task and resource requirements.

The DAM algorithm forms sensor aggregates for target monitoring task. The protocol uses a decision predicate P for each node and a message exchange scheme M that allow the node to decide whether it should participate in an aggregate and how the grouping predicate is applied to nodes. A node determines whether it will belong to an aggregate after it applies the predicate to the data of the node and considers information from other nodes. This algorithm does not consider movement of target.

The EBAM algorithm estimates the energy level by computing the target signal impact area and combines a weighted form of the detected target energy at each impacted sensor. However ambiguous situations may arise regarding target counting.

EMLAM not only estimates positions of target and signal energy using received signals, but also uses the resulting estimates that predict how signals from targets may be mixed at each sensor. Iteration is required to arrive at a good estimate.

Overall the performances of the three algorithms show that the protocols are amenable to implementation on resource-constrained sensor nodes with limited hardware.

4.4.3.3 Synopsis Diffusion

Synopsis diffusion [29] describes a general framework that combines energy-efficient multipath routing schemes with techniques that avoid double counting and achieve significantly more accurate and reliable answers to queries.

The scheme decouples aggregation from message routing by using *order-and duplicate-insensitive (ODI) synopses* and duplicate-insensitive in-network aggregation is used. Synopsis is the partial result generated at a node. Three functions are defined for aggregate computation.

The *synopsis generation* (SG()) function takes sensor readings along with metadata and generates a synopsis. *Synopsis fusion* (SF()) takes two synopses for generating a new synopsis. *Synopsis evaluation* (SE()) is responsible for translating the synopsis into a final answer. The details of the above functions depend on the corresponding aggregate query.

In the *distribution phase* of Synopsis Diffusion algorithm, the aggregate query is flooded in the network and an aggregation topology is constructed. During the next *aggregation phase*, the aggregate values are routed towards the querying node. In this phase each node uses the functions SG() and SF() and finally the querying node uses function SE() to translate into the final answer. The quality of the answer depends on the failure-free propagation path from a node to the query node. This is because with in-network aggregation, sensor readings from nodes will typically be represented in the partial results and in the final result.

However, the main challenge of Synopsis Diffusion is to support aggregates correctly due to the presence of all possible multiple-path propagation schemes. The *order- and duplicate-insensitive (ODI) synopses* that evolve ensure that the final result is independent of the underlying topology. This means that the result is irrespective of both the order and number of times a sensor reading is used in the multi-path routing. There is no mentionable additional power consumption with respect to other existing implementations. Thus this scheme is energy-efficient and highly accurate.

4.4.3.4 TinyDB

Sensor networks are required for obtaining different types of data, based on corresponding applications. Hence it would be beneficial to have a query-processor-interface to sensor networks. At the same time it must be noted that this should consider the resource constraints and power limitations of sensor nodes and the sensor network as a whole. One of the widely used query processors for sensor networks is *TinyDB* [27].

Generally, query processing in sensor networks is viewed as energy-constrained version of traditional query processing. This may be typically implemented by aggregation and filtering operations in the network that help in minimizing communication. *TinyDB* proposes a novel query processing technique, *acquisitional query processing* (ACQP). The focus of ACQP is on the locations and costs of data acquisition at all levels of query optimization, query dissemination, and query processing that would minimize energy consumption.

In TinyDB, a table is maintained that logically contains one row per node per instant in time, with one column per attribute. These attributes are sensor data like light and temperature levels that the particular devices sense and send. Records in the table are acquired only when needed and according to the query raised. Records may be stored for a short time and delivered as and when required. The table is physically partitioned across all sensor nodes in the network. Hence each node stores its own readings in its fragment of the table. If a comparison or other computation of all data is required, these readings from different nodes must be accumulated at some common node for the prescribed task. Queries in TinyDB follow SQL-like constructs and semantics.

Each node in TinyDB maintains metadata that describes its local attributes, events, and user-defined functions. Queries in TinyDB originating at the base station are disseminated into the sensor network. Before dissemination, query optimization takes place. The metadata maintained at each node are periodically copied to the root of the network. The query optimizer uses the metadata. During the broadcast of the query, a node has to decide whether the query applies to itself and needs to be broadcast to its children. This decision is taken if there is a non-zero probability that the node will generate a result for the query. Hence if a node is not able to generate an answer for the query, then the entire subtree rooted at this node is excluded from the query, thus saving the costs of dissemination and execution of queries and corresponding forwarding of results. This is, therefore, an ACQP-related decision.

TinyDB proposes the *semantic routing tree* (SRT) design that allows each node to efficiently determine whether any of its children need to participate in a given query over some constant attribute. In SRT, unlike general routing trees in which link quality plays an important role, choice of parent includes consideration of semantic properties also. The idea is whenever a query Q

with a predicate over P arrives at a node i, i checks to see whether any of the values of P of its children overlaps the query range of P in Q. If the answer is positive, i forwards the query and then receives result. If there is no overlap, Q is not forwarded. However, if Q applies locally, irrespective of the children of i, i executes the query. Finally, if Q does not apply at i or at any of its children, it is simply forgotten. Thus the queries are instantiated and executed in the network.

4.5   Remarks

In spite of large number of energy-aware routing protocols and energy-enabling schemes, there is still need for prolonging network lifetime. Hence different methodologies are explored. One of the most important methods is to *harvest* energy from ambient environmental energy sources [2].

There are a large number of common sources of energy. Examples include light energy, thermal energy and vibration energy. Popular wireless sensor networks research areas mainly focus on light and thermal energy. Generally, in any energy harvesting system, energy may be generated from motion or some thermal or photoelectric source or magnetic activity. The next steps would be to capture and store that generated energy somehow, so that sensors could be driven by that energy. This scenario applies to any wireless sensor network application including industry applications. With a variety of sources of energy, there are different conversion devices for energy harvesting.

4.6   Summary

This chapter discusses various approaches to energy conservation in wireless sensor networks. The authors considered static wireless sensor networks and hence the discussions on energy conservation do not include mobility-based energy conservation schemes.

However, energy management is not complete until integration of different enabling techniques and routing techniques is achieved. The requirement is a single off-the-shelf workable solution. This involves characterizing interactions among different protocols and thereby exploiting cross-layer interactions also. Research is ongoing, since none of the areas described in this chapter seems to have been fully exploited, thus leaving room for further development.

Bibliography

[1] http://en.wikipedia.org/wiki/MovingPictureExpertsGroup.

[2] https://www.tyndall.ie/content/energy-harvesting-wireless-sensor-networks-0.

[3] Jamal N. Al-Karaki, Raza Ul-Mustafa, and Ahmed E. Kamal. Data aggregation in wireless sensor networks: Exact and approximate algorithms. In *Proceedings of IEEE Workshop on High Performance Switching and Routing*. IEEE, 2004.

[4] C. Alippi, G. Anastasi, C. Galperti, F. Mancini, and M. Roveri. Adaptive sampling for energy conservation in wireless sensor networks for snow monitoring applications. In *Proc. of IEEE International Workshop on Mobile Ad–hoc and Sensor Systems for Global and Homeland Security*. IEEE, 2007.

[5] G. Anastasi, M. Conti, M. D. Francesco, and A. Passarella. Energy conservation in wireless sensor networks: a survey. *Ad hoc Networks*, 7(3):537–568, 2009.

[6] G. Anastasi, M. Conti, M. Di Francesco, and A. Passarella. An adaptive and low-latency power management protocol for wireless sensor networks. In *Proc. of 4th ACM International Workshop on Mobility Management and Wireless Access*, 2006.

[7] P. Buonadonna, D. Gay, J. Hellerstein, W. Hong, and S. Madden. Task: Sensor network in a box. In *Proc. European Workshop on Sensor Networks*, 2005.

[8] A. Cerpa and D. Estrin. Ascent: Adaptive self-configuring sensor network topologies. In *Proc. IEEE INFOCOM*. IEEE, 2002.

[9] B. Chen, K. Jamieson, H. Balakrishnan, and R. Morris. Span: An energy-efficient coordination algorithm for topology maintenance in ad hoc wireless networks. *ACM Wireless Networks*, 8(5), 2002.

[10] D. Chu, A. Deshpande, J.M. Hellerstein, and W. Hong. Approximate data collection in sensor networks using probabilistic models. In *Proc. of the 22nd International Conference on Data Engineering*. IEEE, 2006.

[11] A. Deshpande, C. Guestrin, S. Madden, J. M. Hellerstein, and W. Hong. Model-driven data acquisition in sensor networks. In *Proc. of 30th International Conference on Very Large Data Bases*, 2004.

[12] A. Ephremides and O. A. Mowafi. Analysis of a hybrid access scheme for buffered user probabilistic time division. *IEEE Transactions on Software Engineering*, 8(1):52–61, 1982.

[13] D. Ganesan, A. Cerpa, W. Ye, Y. Yu, J. Zhao, and D. Estrin. Geographic random forwarding (geraf) for ad hoc and sensor networks: Energy and latency performance. *Journal of Parallel and Distributed Computing*, 64:799–814, 2004.

[14] B. Gedik, L. Liu, and P. S. Yu. Asap: An adaptive sampling approach to data collection in sensor networks. *IEEE Transactions on Parallel Distributed Systems*, 18(12), 2007.

[15] S. Goel and T. Imielinski. Prediction-based monitoring in sensor networks: Taking lessons from mpeg. *ACM Computer Communication Review*, 31(5), 2001.

[16] S. Goel, A. Passarella, and T. Imielinski. Using buddies to live longer in a boring world. In *Proc. IEEE International Workshop on Sensor Networks and Systems for Pervasive Computing*. IEEE, 2006.

[17] G. Grimmett. *Percolation*. Springer Verlag, 1996.

[18] J. Hill, R. Szewczyk, A. Woo, S. Hollar, D.E. Culler, and K.S.J.Pister. System architecture directions for networked sensors. In *Proc. of ASP-LOS*, 2000.

[19] J. Hill, R. Szewczyk, A. Woo, S. Hollar, D.E. Culler, and K.S.J.Pister. IEEE 802.15.4, part 15.4: Wireless medium access control (MAC) and physical layer (PHY) specifications for low-rate wireless personal area networks (LR-WPANS). 2003.

[20] B. Hohlt, L. Doherty, and E. Brewer. Flexible power scheduling for sensor networks. In *Proc. ACM Workshop on Information Processing in Sensor Networks*, 2004.

[21] B. Kanagal and A. Deshpande. Online filtering, smoothing and probabilistic modeling of streaming data. In *Proc. of 24th International Conference on Data Engineering*. IEEE, 2008.

[22] A. Keshavarzian, H. Lee, and L. Venkatraman. Wakeup scheduling in wireless sensor networks. In *Proc. MobiHoc*. ACM, 2006.

[23] T. Kijewski-Correa, M. Haenggi, and P. Antsaklis. Wireless sensor networks for structural health monitoring: A multi-scale approach. In *Proc. of ASCE Structures Congress*, 2006.

[24] Q. Li, J. Aslam, and D. Rus. Hierarchical power-aware routing in sensor networks. In *Proceedings of DIMACS Workshop on Pervasive Networking*, 2001.

[25] S. Lindsey and C. Raghavendra. Pegasis: Power-efficient gathering in sensor information systems. In *IEEE Aerospace Conference Proceedings, Vol.3, No.16*. IEEE, 2002.

[26] G. Lu, B. Krishnamachari, and C.S. Raghavendra. An adaptive energy-efficient and low-latency MAC for data gathering in wireless sensor networks. In *Proc. of 18th International Parallel and Distributed Processing Symposium*. IEEE, 2004.

[27] S. Madden, M. J. Franklin, J. M. Hellerstein, and W. Hong. TinyDB: an acquisitional query processing system for sensor networks. *ACM Transactions on Database Systems*, 30(1):122–173, 2005.

[28] J.N.Al-Karaki and A.E.Kamal. On the correlated data gathering problem in wireless sensor networks. In *Proceedings of Ninth IEEE Symposium on Computers and Communications*. IEEE, 2004.

[29] S. Nath, P. B. Gibbons, S. Seshan, and Z. Anderson. Synopsis diffusion for robust aggregation in sensor networks. *ACM Transactions on Sensor Networks*, pages 1–37, September 2007.

[30] P. Padhy, R. K. Dash, K. Martinez, and N. R. Jennings. A utility-based sensing and communication model for a glacial sensor network. In *Proc. of Fifth International Joint Conference on Autonomous Agents and Multiagent Systems*. IEEE, 2006.

[31] E. S. Page. Continuous inspection scheme. *Biometrika*, 41(1–2):100–115, 1954.

[32] V. Paruchuri, S. Basavaraju, R. Kannan, and S. Iyengar. Random asynchronous wakeup protocol for sensor networks. In *Proc. IEEE Intl Conf. On Broadband Networks*, 2004.

[33] J. Polastre, J. Hill, and D. Culler. Versatile low power media access for sensor networks. In *Proc. of Second ACM Conference on Embedded Networked Sensor Systems*, 2004.

[34] G. Pottie and W. Kaiser. Wireless integrated network sensors. *Communication of ACM*, pages 51–58, 2000.

[35] M. Pramanik, P. Basak, C. Chowdhury, and S. Neogy. Analysis of energy efficient wireless sensor networks routing schemes. In *Proceedings of 2014 Fourth International Conference on Emerging Applications of Information Technology (EAIT 2014)*. IEEE, 2014.

[36] Q.Fang, F.Zhao, and L.Guibas. Lightweight sensing and communication protocols for target enumeration and aggregation. In *Proceedings of 4th ACM International Symposium on Mobile Ad Hoc Networking and Computing*. ACM, 2003.

[37] V. Raghunathan, C. Schurghers, S. Park, and M. Srivastava. Energy-aware wireless microsensor networks. In *IEEE Signal Processing Magazine, Vol.19, Issue 2*. IEEE, 2002.

[38] M. Rahimi, M. Hansen, W.J. Kaiser, G.S. Sukhatme, and D. Estrin. Adaptive sampling for environmental field estimation using robotic sensors. In *Proc. of IEEE RSJ International Conference on Intelligent Robots and Systems*. IEEE, 2005.

[39] V. Rajendran, J. Garcia-Luna-Aceves, and K. Obraczka. Efficient application-aware medium access for sensor networks. In *Proc. of 2nd IEEE Conference on Mobile Ad-hoc and Sensor Systems*, 2005.

[40] V. Rajendran, K. Obraczka, and J. Garcia-Luna-Aceves. Energy-efficient collision-free medium access control for sensor networks. In *Proc. of SenSys*, 2003.

[41] I. Rhee, A. Warrier, M. Aia, and J. Min. Z-mac: a hybrid mac for wireless sensor networks. In *Proc. ACM SenSys 2005*. ACM–IEEE, 2005.

[42] C. Schurgers, V. Tsiatsis, S. Ganeriwal, and M. B. Srivastava. Optimizing sensor networks in the energy–latency–density design space. *IEEE Transactions on Mobile Computing*, 1(1):70–80, 2002.

[43] C. Schurgers, V. Tsiatsis, and M. B. Srivastava. Stem: Topology management for energy efficient sensor networks. In *Proc. IEEE Aerospace Conference*. IEEE, 2002.

[44] D. R. Stinson. *Combinatorial Designs: Construction and Analysis*. Springer Verlag, 2003.

[45] C. Tang and C. S. Raghavendra. *Wireless Sensor Networks, Compression Techniques for Wireless Sensor Networks*. Springer US, 2004.

[46] Y. Tseng, C. Hsu, and T. Hsieh. Power saving protocols for IEEE 802.11 ad hoc networks. In *Proc. IEEE Infocom*, 2002.

[47] Y.-C. Tseng, Y.C. Wang, K.-Y. Cheng, and Y.-Y. Hsieh. imouse: An integrated mobile surveillance and wireless sensor system. *IEEE Computer*, 40(6):60–66, 2007.

[48] D. Tulone and S. Madden. An energy-efficient querying framework in sensor networks for detecting node similarities. In *Proc. of 9th International ACM Symposium on Modeling, Analysis and Simulation of Wireless and Mobile Systems*. ACM, 2006.

[49] D. Tulone and S. Madden. Paq: Time series forecasting for approximate query answering in sensor networks. In *Proc. of 3rd European Conference on Wireless Sensor Networks*, 2006.

[50] T. v. Dam and K. Langendoen. An adaptive energy-efficient MAC protocol for wireless sensor networks. In *Proc. of ACM Conference on Embedded Networked Sensor Systems*, 2003.

[51] M. C. Vuran and I. F. Akyildiz. Spatial correlation-based collaborative medium access control in wireless sensor networks. *IEEE ACM Transactions on Networking*, 14(2):316–329, 2006.

[52] R. Willett, A. Martin, and R. Nowak. Backcasting: adaptive sampling for sensor networks. In *Proc. of Third International Symposium on Information Processing in Sensor Networks*. IEEE, 2004.

[53] Y. Xu, J. Heidemann, and D. Estrin. Geography-informed energy conservation for ad hoc. In *Proc. ACM MobiCom*. IEEE, 2001.

[54] Y. Xu, J. Heidemann, and D. Estrin. Geography-informed energy conservation for ad-hoc routing. In *Proceedings of Seventh Annual ACM/IEEE International Conference on Mobile Computing and Networking 2001*. IEEE, 2001.

[55] X. Yang and N. Vaidya. A wakeup scheme for sensor networks: Achieving balance between energy saving and end-to-end delay. In *Proc. of IEEE Real-Time and Embedded Technology and Applications Symposium*. IEEE, 2004.

[56] W. Ye, J. Heidemann, and D. Estrin. Medium access control with coordinated adaptive sleeping for wireless sensor networks. *IEEE–ACM Transactions on Networking*, 12(3):493–506, 2004.

[57] Y. Yu, D. Estrin, and R. Govindan. Geographical and energy-aware routing: A recursive data dissemination protocol for wireless sensor networks. In *UCLA Computer Science Department Technical Report, UCLA-CSD TR-01-0023*. UCLA, 2001.

[58] J. Zhou and D. De Roure. Floodnet: Coupling adaptive sampling with energy aware routing in a flood warning system. *Journal of Computer Science and Technology*, 22(1):121–130, 2007.

[59] M. Zorzi and R. R. Rao. Geographic random forwarding (geraf) for ad hoc and sensor networks: Energy and latency performance. *IEEE Transactions on Mobile Computing*, 2(4):349–365, 2003.

[60] M. Zorzi and R. R. Rao. Geographic random forwarding (geraf) for ad hoc and sensor networks: Multihop performance. *IEEE Transactions on Mobile Computing*, 2(4):337–348, 2003.

5

Security in WSN

5.1 Introduction

Wireless sensor networks (WSNs) have earned popularity and acceptance at a very fast pace. Unlike traditional networks, a WSN has design and implementation constraints. Still, resource constraints like storage, power, and computational memory of WSNs did not prove deterrent to their implementation. In fact, this development has happened faster, sometimes leaving a host of issues unresolved. Despite this, WSNs are widely deployed because of their potential low cost solutions with negligible infrastructure to real-life applications. Currently WSNs are deployed for terrestrial [6], underground [4] [57] and underwater [5] [40] applications.

These different scenarios do not leave a single environment in which one is unable to find WSN application. Since each such scenario is unique in nature, it is clear that design and implementation of WSNs are application-specific.

As is well known, sensor networks are utilized for collecting sensor data from sensor nodes dispersed over the region of interest of the application. Specific applications of WSNs range from indoor deployment scenarios in homes, offices, and hospitals to outdoor scenarios like forests, flood-prone areas, earthquake-prone areas, and battlefields. Almost all of these applications work with real-time data where timeliness of data is an important issue. Likewise in all applications, correctness of data plays an important role [81]. Sensed data are not only collected, but depending upon the application, are sometimes aggregated also, before being sent to the sink or the base station. There may be a long route from the source of the data to its ultimate destination. Hence care must be taken to maintain correctness and freshness of data during this long-haul path. The path may have numerous forwarder nodes. The need is to ensure that these nodes forward data timely and correctly. Thus various issues need to be considered for successful execution of WSN applications.

5.2   Challenges in WSN

The inherent nature of the sensor network and its constituent sensor nodes including sinks and base stations present many research challenges to the designers and implementers. The main challenge is to provide the capability of self-organization. This is required from the perspectives of both network maintenance and management as well as application scenarios. Other challenges stem mainly from the resource constraints of the nodes in the network. Wireless communication among the nodes is another major contributor to the challenge. This is because wireless communication allows vulnerability to unwanted and unauthorized access that may usher in associated security threats. Tiny sensor nodes add to physical security threats. Traditional security measures may not be successful in WSNs due to these constraints.

The major constraints in WSN are identified as follows [90]:

1. Limited memory - Since sensor devices are tiny, a very small amount of memory space (like 10K RAM, 48K program memory in common TelosB motes) is available.

2. Limited storage space - Storage space is very limited in tiny sensor devices (like, 1024K flash storage in common TelosB motes).

3. Limited power Once deployed it may not be possible to replace or recharge sensor nodes.

4. Unreliable communication - Generally, packet-based routing protocols are inherently unreliable due to broadcast mode of communication, channel errors and congestion. These also add up to greater latency in the network.

5. Unattended operation - Sensor nodes are generally left unattended because the region of their deployment is unreachable. Thus sensor nodes may suffer from poor weather conditions and physical attacks. The conditions of the nodes and the attacks are hard to detect since they are generally remotely managed and maintained.

It is logical to say that traditional security measures like cryptographic functionalities including integrity checking and other security-related tasks like exchanging security information and keys, storage of such data would take up much storage space and require complex computation and memory space. All of these high overheads for resource create constrained WSNs.

WSNs as networks have all the security requirements of a typical network and may have a few additional requirements since sensor nodes are special types of devices. However, a typical WSN application may or may not require all of these. Hence we have to consider all the following possible security requirements of sensor networks.

1. Data confidentiality - This is undoubtedly the most important security requirement of any network, more so because sensor networks are used in a wide variety of military applications and no one wants sensor data to be leaked to unauthorized entities. Even public information transmitted over such network must not also fall into unauthorized hands. Confidentiality is generally provided by cryptographic techniques in such a way that only a valid receiver is able to decrypt data that is encrypted by the sender.

2. Data integrity - Due to reasons like unfavourable environmental conditions or presence of malicious nodes in a network, correct data may not reach its destination. During rough weather conditions communication may fail in the middle of a transmission, leaving the receiver with incomplete or incorrect data. Malicious nodes may sneak into a data packet and change some bits and then again send it to the original destination. So, it has to be somehow ensured that the data received by the receiver is correct and this creates the need for data validation at the receiver. Providing data integrity methods would therefore ensure that data is not changed during transit.

3. Timeliness of data - It is typical for an application to always demand fresh data. Particularly if the application is a real-time one or information to be transmitted is related to security issues like key exchange, data must be delivered on time as late arrival may waste resources or cause fatalities. However, in the absence of any clock or time synchronization methods (which may be prohibitively costly in case of WSNs), using none may be a feasible alternative.

4. Availability - This has always been one of the key aspects of security. Providing security to applications always burdens the application with additional computation, communication and storage needs. Hence the original goal of having the system or sensor network always available may be defeated by the effort of securing itself, since security will add to the complexities. The normal operations of the network may be hampered due to the implementation of security-related issues, thereby impacting availability. What is therefore required is a fine tuning of what is required for a particular application.

There are a host of other security requirements [76] [90] that may not be primary, but these are also important and deserve mention. Few of these are regarding general issues whereas others may be specific to applications. These security requirements are detailed as follows:

1. Secure routing protocols - Threats to compromise sensor node, data and communication channel loom large over WSNs. Since WSNs are distributed in nature, communication is a major issue. Data and information transmission take place continually. Hence it is of

utmost importance to provide adequate security during transmissions. Routing protocols must be armed with proper security measures. Because of the special nature of WSNs, their routing protocols are also different from those used in other networks. Most routing protocols in WSNs are multi-hop in nature and must deal with intermediate forwarder nodes. Thus the entire path from source to destination has to be secured to ensure that data is not tampered with in transit and delivery is on time.

2. Self-organization - Wireless sensor networks are typically ad hoc in nature with no central management. The nodes therefore have to adapt to situations based on somewhat limited information they gather from time to time. This feature is fraught with security risks, since there are few ways of distinguishing between genuine and faulty or fake information. Hence mechanisms like fault tolerance may be adopted so that a node is able to self-organize itself accordingly.

3. Secure localization - Localization is the ability to accurately locate a sensor node in the WSN. This is one of the most important requirements for many WSN applications. For a network to perform self-organization, localization may be a prerequisite. WSN is distributive in nature and location information of a sensor node may be calculated based on the location information of other nodes in the network. Thus location information needs to be secured lest a malicious node finds it and starts manipulating by reporting false data. Secure location information will help to find faulty nodes accurately and thereby aid in self-organization.

5.3 Attacks in WSN

Sensor networks are generally deployed in outdoor environments. The distributed nature of deployment and an unattended network make any WSN vulnerable to attacks. Furthermore, open environments are also prone to hostilities of nature and other threats that are common in networked environments. A sensor network may face several types of attacks that need to be addressed and measures must be taken to counter these attacks.

5.3.1 Attack Categorization

Attacks in sensor networks can be classified into the following categories [94]:

1. Outsider versus insider attacks - Outsider attacks to a WSN are launched from nodes that do not belong to that WSN. The outsider

attack generally targets confidentiality and authentication mechanisms of the system [84]. On the other hand, insider attacks occur when the nodes of a WSN behave in a manner unexpected of them.

2. Passive versus active attacks - Passive attacks mean attackers do not directly get involved with the system but resort to behaviours like eavesdropping or monitoring (only) packets exchanged within a WSN, whereas active attacks involve not only monitoring but also modifications of data or creation of fake data. Both types attempt to play with the confidentiality and authentication of the WSN [84]. Further, active attacks may stealthily compromise network service integrity by making it accept false data [84].

3. Mote-class versus laptop-class attacks - In mote-class attacks, an adversary may compromise a couple of nodes or introduce intruder nodes similar in capabilities to the WSN nodes for launching attacks. In laptop-class attacks, an adversary tries to use devices such as laptops and smart devices whose capabilities exceed those of a WSN node to attack a WSN. The capabilities of these devices include far-reaching transmission range, processing power, and energy reserves (including battery back-up) that are not present in WSN nodes.

Any of these above categories of attacks may subsequently lead to an attack on network service availability. Denial of service (DoS) occurs when a situation obstructs or disrupts or eliminates the normal activities of the network. In fact, any kind of attempt of an adversary to disrupt, subvert, or destroy the network or its capacity to perform its expected function and achieve desired performance level, can be termed a DoS attack [72]. DoS attacks may target any layer of a sensor network, as elaborated later in this section.

There may be physical attacks and remote attacks. Attacks by remote adversaries affect different networking layers of the WSN.

5.3.1.1 Physical Attack

In this type of attack, it is possible to gain full control of a sensor node through direct physical access. It may not be practical to build a sensor node with tamper-proof material since it will be costly. But if an adversary is able to compromise a node, it can obtain information like cryptographic key secrets or even the code stored in the memory. The attack can also destroy a node physically by writing into its physical memory. The attacker may then gain unrestricted access to the network. The normal operation of the node may not be hampered while this attack is underway. Sometimes physical attack may require the node to be physically removed from the deployment area. This, however, will not go unnoticed as neighbours will be able to detect the absence, but may not be able to find or fix the cause.

5.3.1.2   Attacks at Different Networking Layers

- Attacks at Physical Layer

  The tasks of the physical layer include data transmission and reception, frequency selection, carrier frequency generation, signaling function and data encryption and looking after the transmission media among communicating nodes [76].

  1. Jamming: WSNs use radio-based shared transmission media. Jamming or radio interference is a common attack phenomenon. This attack is very easy to implement since the adversary only has to know the wireless transmission frequency used in the WSN. The attacker will go on transmitting radio signal randomly at the same frequency. Obviously, this will interfere with the signals sent by other nodes and hence receivers (those within the range of the attacker) will not be able to receive the signal meant for them. This will continue as long as the jamming signal continues and the affected nodes effectively are isolated in the network. Several techniques are proposed that prevent jamming. Frequency hopping spread spectrum technique for countering jamming [52] may not be effective due to issues like processing overhead and limited range of frequencies for WSN. Another proposal for anti-jamming is ultra wide band transmission [24]. UWB transmission is based on sending very short pulses in the order of nanoseconds across a wide frequency band. Hence it is very difficult to detect. This technique is suitable for WSN because energy consumption is low.

- Attacks at Data Link Layer

  The functionalities of the data link layer include multiplexing of data streams, detection of data frames, medium access, and error control [6]. Setting up reliable point-to-point and point-to-multi-point connections is one of the tasks of the data link layer. Common attacks in this layer are purposefully generated collisions and resource exhaustion.

  1. Collisions: Collisions occur when multiple nodes try to transmit using the same frequency simultaneously. An attacker may target certain types of packets and generate collisions. Colliding packets generally result in checksum mismatches at the receiver. Thus the sender node has to retransmit the message. In this way, the energy of the node will be depleted. Using error-correcting codes is the most typical way to handle such collisions. Error-correcting codes are effective up to a certain

level of collision, for example, a collision caused by environmental or probabilistic errors. But processing and communication overhead will increase, so, correcting codes may not be too effective in WSNs.

2. Resource exhaustion: Repeated collisions cause exhaustion of resources because sender nodes will go on resending packets that collided. If these unnecessary transmissions cannot be stopped, energy of the transmitting nodes and the nodes along the path to the receiver will decrease at an alarming rate. Rate limits to MAC admission control may be used so that the network ignores excessive requests. This in turn will prevent energy consumption by repeated transmissions [94]. Time-division multiplexing may also be used where each node is allotted a time slot for transmission [94]. However collisions still cannot be avoided.

3. Unfairness: The attacker may resort to the above-mentioned link layer attacks randomly in a way such that nodes may miss transmission deadlines. An attacker always tries to degrade the performance of the network that may ultimately sometime lead to usurpation.

- Attacks at Network Layer

The main task of the network layer is routing. The network layer in a WSN must also consider that WSN is a data-centric, location-aware network with attribute-based addressing.

1. Attacking Routing Information: Routing information is the most important component of routing in a network. If routing information is violated, an attack may be executed directly in the network layer. Such attacks would obstruct flow of traffic in the network. Disruption in normal flow may be caused by various means like spoofing, altering or even replaying routing information [94]. Disruptions may result in routing loops, altering routes and also partitioning of the network and an increase in latency. Message authentication code (MAC) may be appended with messages whereby receivers can detect whether a message has been altered or not. Replayed information is generally identified by time stamps or sequence numbers.

2. Sinkhole attack: The compromised node, somehow manipulates its routing metric such that other nodes rush to route their data through it. Thus, the compromised node is now able to launch any type of attacks like selective forwarding, modifying data that passes through it and the more severe blackhole attack.

3. Blackhole attack: The nodes in a network that drop or discard data (passing through them) are referred to as blackholes

and they can launch blackhole attack. Usually it is not easy to detect these nodes. These nodes may only be detected by monitoring lost traffic.

4. Wormhole attack: An adversary connects any two points (probably far away from each other) in the network using a direct low-latency communication link. This newly formed link, called a wormhole link, can be established by using ethernet cable, long-range wireless transmission or an optical link. With establishment of this link, the adversary captures data (packets) on one end, tunnels them through the wormhole link and replays them at the other end. The wormhole attack is possible even when the attacker does not compromise any node and even if communication is authenticated and confidential. Hu et al. [43] presented a mechanism called packet leashes for detecting and defending against wormhole attacks. The proposed mechanisms are also suitable for WSNs.

5. Sybil attack: The attacker creates more than one identity (pseudonym) of a node. This sybil attack thus fools a node into believing that redundancy has been achieved for schemes that require it and other schemes required for network maintenance, fault tolerance and distributed storage.

Besides the above attacks, a node may be compromised in such a way that it starts forwarding selectively and simply dropping data. In other situations, nodes may be tricked into believing that they have a certain neighbourhood area whereas the reality is that an attacker node may be using high-powered transmitters and may broadcast false superior routes to a base station. It is also possible for an attacker to spoof acknowledgement by overhearing conversations. This attack may take a serious form since (spoofed) acknowledgements may come from a node that is actually dead.

- Attacks at Transport Layer

  The transport layer maintains end-to-end connections. However, a couple of attacks are prevalent in this layer also.

  1. Flooding: Transport layer protocols need to maintain states at both communicating ends and resources are required. An attacker may continue requesting new connections until resources are exhausted or some maximum limit is reached and further legitimate requests could not be met. However, it is unlikely that an attacker would have unlimited resources to continue this attack forever.

  2. Desynchronization: The attacker changes the sequence numbers

of packets to disrupt the communication protocol. Authentication of packets may be a possible solution.

- Attacks at Application Layer

  The basic task of the application layer is management of the WSN including the information or data collected at base stations.

  1. Attack by the application: An application may generate a large number of messages such as control and alert conveyed to deployed nodes, thereby generating huge traffic in the network. Limits may be put on the number of alert messages or filtering may be applied to such messages after checking some parameters for validity.
  2. Programming attack: Deployed nodes may be reprogrammed in extreme cases. This may be done by sending false programs to nodes. This type of attack may be countered by checking integrity of the received program.
  3. Path-based denial of service attack: The path frequently used by nodes for data packets to reach base stations may be used for sending large number of bogus data packets. The nodes will always remain busy and sometimes exhaust their resources in forwarding these packets, thereby denying legitimate packet traffic.

Different resources in WSN at different layers may be targeted by an attacker such that proper functioning of legitimate nodes is affected, and in worst cases, stops. Jamming and node tampering are two well known DoS attacks in the physical layer. In the link layer, collision, battery exhaustion, and unfairness are DoS attacks. In the routing layer, DoS attacks include spoofing, replaying, misdirection of traffic, hello flood attacks, and homing. Flooding and desynchronization are the two common attacks in the transport layer. Path-based DoS (PDoS) attacks are application layer DoS attacks. Reprogramming of the WSN is also an application layer DoS attack.

5.4 Protection against Attacks

Attacks in WSN can take place in any layer. The nature of attacks determines the severity. Some attacks may modify the transmitted data which can be routing information or control information while others may compromise the nodes. Thus there is need for securing in-transit data and authentication of nodes. Security of in-transit data involves confidentiality as well as integrity.

Providing confidentiality will not permit attackers to read or modify data
whereas integrity check at the receiving end will validate the data. Authentication of nodes will help distinguish between compromised and uncompro-
mised nodes. Security in WSN requires choice of proper cryptography. This is
particularly true for WSNs because of the constraints and hence the adopted
cryptographic methods should be evaluated for size (code, data), processing
time and complexity and power consumption.

5.4.1   Cryptography in WSN

Cryptography is the ancient art of writing in secret code. Cryptography is
used not only to protect data by hiding it from unintended users but also for
user authentication and maintaining data integrity. Cryptography is classified
into two major categories: public key or asymmetric cryptography and private
key or symmetric cryptography.

5.4.1.1   Public Key Cryptography

Public key cryptography methods use two types of keys: public key and secret
or private key. Each node keeps a pair of these keys that are mathematically
related to each other. The public key of a node is known to all others in the
network whereas the private key is known only to itself. It is not feasible to
compute a private key even if the corresponding public key is known.

Traditional and widely used schemes for public key cryptography include
Diffie-Hellman [27], RSA [80], El Gamal [36] and others. Since each of these
was meant to be used by and usually implemented in large-scale platforms,
they require considerable resources. Wireless and mobile devices like cellular
phones generally require fewer computational resources. Sensor devices are at
a notch lower than these devices. Hence use of these algorithms is restricted in
WSN. Developers are seeking algorithms more efficient than these. Algorithms
based on elliptic curve cryptography (ECC) such as elliptic curve Diffie Hellman (EC-DH) and elliptic curve-digital signature algorithm (EC-DSA) [96]
execute considerably faster while maintaining the same levels of security as
those of the traditional algorithms.

Public key algorithms generally execute thousands of multiplication instructions to perform a single security-related operation. The efficiency of
a microprocessor execution of a public key algorithm is primarily governed
by the number of clock cycles required to perform one multiply instruction
[18]. This means that in constrained wireless devices RSA-like algorithms may
require tens of seconds to perform encryption and decryption operations. Energy requirements for such an operation may be thousands of nano joules
for a 128-bit result [18]. Hence focus shifted toward ECC at first. For a far
smaller key size, ECC appears to offer equal security with less processing and
communication complexity. For example, RSA with 1024-bit keys (RSA-1024)
is equivalent in strength to ECC with 160-bit keys (ECC-160)[33]. Another

TABLE 5.1

Public key cryptography: average ECC and RSA execution times [39]

Algorithm	Operation time (seconds)
ECC secp160r1 [40]	0.81
ECC secp224r1 [40]	2.19
RSA-1024 public key $e = (2)^{16} + 1$	0.43
RSA-1024 private key w. Chinese remainder theorem	10.99
RSA-2048 public key $e = (2)^{16} + 1$	1.94
RSA-2048 private key w. Chinese remainder theorem	83.26

recommendation from RSA Security is RSA-2048 as the new minimum key size which is equivalent to ECC with 224 bit keys (ECC-224) [46] [75]. Table 5.1 [94] summarizes the execution times of ECC and RSA implementations on an Atmel ATmega128 processor (used by Mica2 mote) [39]. The execution time is measured on average for a point multiplication in ECC and a modular exponential operation in RSA.

It is evident from Table 5.1 that using a small public key $e = (2)^{16} + 1$ RSA public key operation is faster than ECC point multiplication, but ECC outperforms RSA private key operation by an order of magnitude.

Table 5.2 [94] shows the energy cost of authentication and key exchange based on RSA and ECC cryptography on an Atmel ATmega128 processor [91]. The key exchange protocol works with a client initiating the communication and a server responding to the initiation. The WSN is assumed to be controlled by a central base station with each sensor node having a certificate signed by the central server's private key using a RSA or ECC signature. After verifying each other's certificates, the session key to be used in the communication is negotiated. As apparent from Table 5.2, ECDSA signatures are significantly cheaper than RSA signatures and ECDSA verifications are comparable with RSA verifications. Though at the client side the energy costs for these two key exchange protocols are practically similar, the ECC-based key exchange protocol is far better than the RSA-based key exchange protocol at the server side.

TABLE 5.2

Public key cryptography: average energy costs of digital signature and key exchange computations (milliJoules) [91]

Algorithm	Signature		Key Exchange	
	Sign	Verification	Server	Client
RSA-1024	304	11.9	15.4	304
ECDSA-160	22.82	45.09	22.3	22.3
RSA-2048	2302.7	53.7	57.2	2302.7
ECDSA-224	61.54	121.98	60.4	60.4

But ECC-based algorithms have to pay a price too in terms of using complex arithmetic primitives and a large number of temporary operands. This is in sharp contrast to RSA or El Gamal, that require only one single arithmetic primitive and few operands. Another drawback faced by ECC is its large storage requirements that render it less scalable and less attractive for energy-efficient implementations.

However, with respect to execution time and energy cost, ECC has performance advantages over RSA as key size increases. TinyPK [95], based on RSA, and TinyECC [1], based on ECC, used Mica2 motes as the implementation platform in TinyOS environment [94]. So it may easily be inferred that public key cryptography is a viable option for implementing security issues in WSNS. But public key operations still remain expensive. This is mainly due to the constraints on computation and power consumption of WSN nodes. Hence a large number of research works on security of WSNs focus on symmetric key cryptography.

5.4.1.2   Symmetric Key Cryptography

In symmetric key cryptography a node (N1) maintains secret or private keys (k1,2, k1,3, ..) for each of its communication partners (N2, N3, ..) respectively. The key (say, k1,2) is the same (hence, symmetric) that N2 also maintains for N1, and so on. The same key is used for encryption at the sender's end and for decryption at the receiver's end. This is in contrast with public key cryptography technique, in which a node has to maintain only one secret or private key with itself, since the other key, that is, the public key may be maintained and available through some well-known mechanisms in the public domain. Thus, a node using symmetric key cryptography, may have to exert extra efforts (reliable storage space, efficient data structures for retrieval, etc.) for maintaining a list of such symmetric keys safely. Moreover, another important concern is the generation (or distribution) of the same key for each communication partner.

The performance of symmetric key cryptography depends on embedded data bus width, which is generally less than the requirements of encryption algorithms and the instruction set architecture (ISA), which affects certain algorithms, since most embedded processors may not support certain operations that directly improve performance of some algorithms. For example, most embedded processors do not support the variable-bit rotation instruction ROL (rotate bits left) of the Intel architecture that improves the performance of RC5.

A number of popular symmetric key encryption schemes, like, RC4 [66], RC5 [78], IDEA [66], SHA-1 [22], and MD5 [66], [77], TEA [97], RC6 [79], Rijndael [23] and MISTY1 [65] were evaluated by different researchers [35], [85], [38], [51]. It was found that hashing algorithms (MD5 and SHA-11) incurred almost an order of a magnitude higher overhead than encryption algorithms (RC4, RC5 and IDEA). Other evaluation results showed that while Rijndael is suitable for high-security and energy-efficiency requirements, MISTY1 is suitable for good storage and energy-efficiency requirements. The evaluation results [51] did not find RC5 and RC6 to be suitable, unlike results in [35] and [85] respectively. The execution time of RC5 (64-bit block using an 80-bit key) on an Atmel ATmega128 processor is 0.26 ms, magnitudes less than public key cryptographic technique timings as mentioned in Table 5.1. The energy cost of SHA-1 (C) [22] is 5.9 mJ/byte and that of AES-128 Enc/Dec (assembly) [23] is 1.62/2.49 mJ/byte. It is therefore evident that symmetric key cryptography is faster and consumes less energy than public key cryptography.

Though symmetric key cryptography techniques may be found to be suitable because of their relatively lower energy costs and faster speed, key distribution schemes are still far from perfect. It is evident that for both schemes of cryptography, key generation and distribution are necessary when using not so secured limited range wireless transmission of sensor nodes. There may not be one single scheme that is suitable for all types of application of WSNs. Various such key distribution and management schemes have been proposed and analyzed over the years, a few of which are described below.

5.5 Key Management

Wireless sensor networks may have a flat topology (similar to the concept of distributed WSN) or hierarchical or clustered topology. Key distribution and management schemes may be based on the specific characteristics of the topologies and also consider the methodology of generation and distribution of keys. There are two approaches, the probabilistic approach and deterministic approach, and a hybrid approach also finds its mention in the literature [16]. Probabilistic approaches use random selection and distribution of key chains from a key pool to sensor nodes, whereas in deterministic approaches, deterministic methods are used to design the key pool and the key chains. Hybrid approaches try to combine both by using probabilistic approaches on deterministic solutions with the aim to improve scalability and resilience. Other classification criteria may be based on the time of deployment, that is, predeployment, dynamic distribution or post-deployment or both, depending on the application and its requirements.

5.5.1 Key Management in Distributed WSN

5.5.1.1 Pair-wise Key Pre-distribution Schemes

Apart from trivial schemes like distributing a single master key or each node storing a unique pairwise key prior to deployment, there exist several other schemes as well. A few of them are described below.

1. The random pairwise key scheme proposed by Chan et al. [19] requires each node to store a random set of Np pairwise keys where p denotes the probability that two nodes are connected. During key setup, a pairwise key generated for each ID pair (of matched nodes) is then stored in both nodes' key chains. During shared-key discovery, each node broadcasts its ID and hence neighboring nodes can tell whether they share a common pairwise key.

2. Closest (location-based) pairwise keys pre-distribution [61] uses pre-determined location information of nodes to find and share pairwise keys with closest neighbours. Both schemes have good key resilience and are scalable.

3. ID based one-way function (IOS) [54] and multiple IOS [54] schemes assume a network to be a regular graph. While the first has good key resilience, the second scheme is more scalable with less key resilience.

5.5.1.2 Master Key-based Key Pre-distribution Schemes

1. Broadcast session key negotiation protocol (BROSK) [49] uses pre-deployed single master key, based on which a communicating node pair generates session key. The scheme does not have good resilience because master key may be compromised.

2. Lightweight key management system [32] considers deployment of nodes in generations and maintenance of group authentication key (gk) and a key generation key (kgk). Nodes belonging to same generation authenticate themselves by gk and then generate session keys using gkg and nonces. Nodes belonging to different generations use both keys besides generating and authenticating each other by some secret corresponding to each new generation.

5.5.1.3 Random Key Chain-based Key Pre-distribution Schemes

1. Basic probabilistic key pre-distribution [34] is based on probabilistic key sharing. A large pool of keys with identities are generated in the key setup phase. For each sensor node, k keys (without replacement), that form the key chain, are randomly drawn from the key pool. In shared-key discovery phase, neighbour nodes exchange and compare list of identities of keys.

2. In cluster key grouping [44] key chains are divided into C clusters with a start key ID for each cluster. Remaining cluster key IDs are implicitly known from the start key ID. In shared-key discovery phase only start key IDs for clusters are broadcast.

3. The scheme proposed in [30] develops a key pre-distribution scheme based on deployment knowledge model where sensor nodes are divided into n groups deployed as resident points arranged in a two-

dimensional grid. The key pool is divided into subsets where (i) two horizontally and vertically neighbouring key pools have some keys in common, (ii) two diagonally neighbouring key pools have some other keys in common, and (iii) non-neighbouring key pools do not share a key.

5.5.1.4 Combinatorial Design-based Key Pre-distribution Schemes

1. The concept of combinatorial design based pairwise key pre-distribution scheme [15] is based on block design techniques of combinatorial design theory.

2. Other similar combinatorial design theory approaches are proposed in [53].

5.5.1.5 Key Matrix-based Dynamic Key Generation Schemes

1. Blom's scheme [8] uses one public matrix and one private matrix. With the public matrix having (+1) linearly independent columns, this scheme is secure, that is, keys are secure if no more than nodes are compromised. The overhead of the scheme involves matrix multiplication and message exchange.

2. Multiple space key pre-distribution [31] improves the resilience of Blom's scheme, whereas scalability of the Blom's scheme is improved in the multiple space Blom's scheme (MBS) [54].

5.5.1.6 Polynomial-based Dynamic Key Generation Schemes

1. Polynomial-based key pre-distribution [10] distributes a polynomial share (partially evaluated polynomial) to each sensor node by using whichever pair of nodes generates a link key.

2. Polynomial-based key pre-distribution [10] is combined with the key-pool idea discussed in [34], [19] and proposed in [60]. It generally improves resilience and scalability by considering that not all pairs of nodes may have to establish keys.

3. Location-based pairwise keys [61] uses location information where a deployment area is divided into R rows and C columns. The scheme uses bivariate polynomials and is based on polynomial-based key pre-distribution [10].

5.5.1.7 Group-wise Key Distribution Schemes

1. In lightweight key management [32] groups of sensor nodes are deployed in different phases. The scheme proposes to distribute group-wise keys through secured links (using pairwise keys).

5.5.2   Key Management in Hierarchical WSN

In hierarchical WSN (HWSN), presence of base stations may make key distribution somewhat easier. It may be assumed that base stations share distinct pairwise keys with each sensor node. These keys can then be used to establish secure links with each other. This gives rise to solutions using pairwise, group-wise and network-wise key distribution in HWSN.

5.5.2.1   Pairwise Key Distribution Schemes

1. Localized encryption and authentication protocol (LEAP) [102] establishes pairwise keys for a sensor node with its immediate neighbour. In the key setup phase, nodes receive a general key K_g. A node uses K_g and a one-way hash function H to generate its master key K_m. In shared key discovery, a neighbour will respond after getting the first node's broadcast information. Both nodes then generate a session key. However, all generated session keys can be compromised once K_g is compromised,

2. ESA [50] uses an approach similar to LEAP, where sensor nodes are grouped into domains supervised by base stations.

3. The authors of SNEP [35] propose that each pair generates encryption keys and MAC keys by using a shared master key.

5.5.2.2   Group-wise Key Distribution Schemes

Group-wise keys are required in hierarchical WSNs for securing multicast communication.

1. ID-STAR [17], as the name suggests, uses identity-based cryptography [83], [12]. Sensor nodes' public keys are derived from their identities. It is also possible to use existing pairwise key structures to establish groups-wise keys.

2. Localized encryption and authentication protocol (LEAP) [102] also provides mechanisms to generate group-wise keys which follows the LEAP pairwise key establishment phase. A node that wants to establish a group key with its neighbours generates a unique group key.

3. In ARMS (authenticated routing message in sensor networks) [56] each pair of neighbouring nodes shares a secret key. The sender generates random key K_1, before actual transmission. Then a short one-way key chain F(K_1), K_1, is derived and F(K_1) (commitment) is sent to all neighbors via authenticated unicast. In the actual packet a new commitment F(K_2) and MAC of the message are provided. The receiver is able to validate this packet since the old commitment is known and subsequently a new commitment F(K_2) is established. Though ARMS requires only 20 additional bytes per

message with low memory requirement (16 bytes for receiver, 48 bytes for sender), generation of 8 bytes of random data by sender per every two messages becomes a costly affair.

5.5.2.3   Network-wise Key Distribution Schemes

Master key-based schemes

Network-wise keys are required to secure traffic from base stations to sensor nodes in hierarchical WSNs.

1. Multi-tiered security [85] considers three different information flows in the network which are mobile code, locations of sensors nodes and application data. Each sensor node obtains an active master key from a preloaded set of master keys and a pseudo random function with a seed. RC6 is used for encryption. Three security levels are defined. However, the preloaded credentials may be compromised.

TESLA-based schemes

1. Timed efficient stream loss-tolerant authentication (TESLA) [73] is a multicast stream authentication protocol using a delayed key disclosure mechanism where the key used to authenticate the ith message is disclosed along with $(i + 1)th$ message.

2. μ-TESLA [35], considered to be the micro version of TESLA, provides authentication for data broadcasts, and requires the base station and sensor nodes to be loosely time synchronized. Basically, a base station randomly selects the last key of a key chain, and applies one-way public function to generate the rest of the chain. Given any key, a node is able to generate the rest of the keys up to that key, but not the next one in sequence. Storage problems may arise since a node has to store a message until the base station discloses the key in the next time slot.

   Another variation of TESLA is described in [11].

5.6   Secure Routing in WSNs

A large number of routing protocols for WSNs have been proposed over the years. The basic objective of routing protocols for ad hoc networks including WSNs had always been the establishment of a route with minimum number of hops or a route that consumes minimum energy. Security in routing was generally not considered to be one of the main objectives. However with the possibility of different types of attacks as discussed in Section 5.3.1.2, it is necessary to consider and incorporate security measures while designing routing protocols for WSNs.

Typically, a secure routing protocol may be needed only for guaranteeing message availability. In traditional networks, other aspects of security like message integrity and confidentiality and sender authenticity are dealt with by higher network layers using end-to-end security mechanisms such as SSH or SSL. This is convenient because messages need not be accessed at intermediate hops in traditional networks. However, in-network processing may be required for some applications in sensor networks. In such situations providing end-to-end security would never serve the purpose [47]. But at the same time it must be mentioned that, given the constraints of sensor nodes, implementation of any security measure is challenging. Communication bandwidth is also prohibitively costly in terms of enegy consumption, for example, consumption of power for each bit transmitted is approximately the power needed for executing more than 8 million instructions [42]. Undoubtedly, power is the scarcest resource in sensor networks. Hence the design of secure routing protocols for WSNs should also consider the factor of energy consumption, among others.

5.6.1 Attacks on Routing Protocols

Any sensor network routing protocol is prone to attacks as mentioned in Section 5.3.1.2. Here we present a brief overview of various attacks on a few routing protocols.

5.6.1.1 Directed Diffusion

Directed diffusion [45] is a well-known data-centric WSN routing algorithm in which base stations flood interest for specific data and set up gradients to fetch information. Sensor nodes satisfying the interest follow the reverse path of interest propagation to disseminate information. The work in [37] proposes a multipath variation of directed diffusion.

One of the possible attacks on this routing protocol can be strong reinforcement to the nodes to which interest is sent by an adversary that wants to handle the resulting flow of information. At the same time the adversary will spoof high-rate low latency events to the nodes from which it receives interest. Another possible attack can be selective forwarding.

The work in [93] describes the design of a secured directed diffusion protocol. The scheme assumes pre-distribution of a shared key between the sink and all nodes in the network along with the first element of pre-deployed one-way hash-chain in the sink. A modified immediate authentication TESLA [73] is used to authenticate various stages of this protocol.

5.6.1.2 LEACH

Low-energy adaptive clustering hierarchy or LEACH [41] organizes sensor nodes into clusters with a cluster-head per cluster. The cluster-head aggregates data sent by its member nodes for transmission to base station. A randomized rotation for choosing cluster-heads is used to evenly distribute energy consumption over all nodes in the network. In the setup phase each node has the ability to probabilistically decide whether to be a cluster-head based on its remaining energy and a globally known desired percentage of cluster-heads. Once cluster-heads are chosen and the remaining nodes join clusters and the allotted corresponding TDMA schedule for communication with respective cluster-heads is established, the steady-state phase begins where nodes send data to the cluster-heads.

A common attack to this protocol may be the hello flood attack which can be used by an adversary to send powerful advertisements during the setup phase. It is possible that nodes will choose the adversary as cluster-head because of its strong signal.

The work in [68] uses random key pre-distribution, to secure communication in hierarchical (cluster-based) protocols like LEACH.

5.6.1.3 Rumor Routing

Rumor routing [13] is a probabilistic routing protocol where nodes try to match queries with events. After observing an event, a node sends an agent on a random walk in the network. Agents carry information about the events and the nodes visited and the next hop in the path to the event. On arrival at a hitherto unvisited node, the agent informs the node about the event it is carrying and also takes information about the event and leaves for the next unvisited hop, provided it is still alive. A base station also disseminates its query in the same way through an agent.The route from a base station to a source is established when a query agent arrives at a node that has an answer to the query (because it has previously been visited by an event agent).

A common attack can be to deny agent forwarding or removing or modifying event or query information from an event or query agent. Selective forwarding attack is also possible [47].

5.6.1.4 Geographic Adaptive Fidelity

Nodes are placed in virtual "grid squares" according to geographic location and expected radio range in GAF or geographic adaptive fidelity [99]. Any two nodes in adjacent grid squares can communicate. Nodes are in one of the following states: sleeping, discovery, active. Active nodes participate in routing while discovery nodes determine whether they will be needed or not. Sleeping nodes have their radios turned off. The objective of GAF is to have only one active node in a grid.

An adversary can ultimately become the only node in its grid by periodically broadcasting high-ranking messages. It can then use the selective forwarding attack or shut itself up. An attacker can target individual grids, thereby putting all nodes in the network in sleeping mode.

5.6.1.5 SPAN

In SPAN [20] nodes decide whether to sleep or join a backbone of coordinators. This backbone attempts to maintain routing fidelity in the network. Coordinators have to stay awake all the time, while other nodes go into power saving mode and periodically send and receive HELLO messages to determine if they should become a coordinator.

An attack may attempt to prevent nodes from becoming coordinators. It may work as follows: the adversary partitions the area into a number of cells: C_1 ; C_2 ; . . . ; C_n. ID_i (in all probability, a bogus coordinator node id) is chosen as coordinator for C_i. The adversary broadcasts n HELLO messages that can be heard by every node announcing the coordinator ID_i and its neighbours, thus making the nodes in a cell to believe their neighbours. Each bogus coordinator also declares its neighbours. The adversary thus effectively disables the entire network since no real nodes now actively participate in routing.

5.6.2 Countermeasures for Attacks

To prevent the above-mentioned attacks and other attacks on different routing protocols, a few countermeasures are considered. This list of countermeasures can never be said to be a complete list since different application scenarios may demand some other appropriate countermeasure.

Attacks caused by outsiders in WSNs can be prevented by link layer encryption and authentication. This may defend the network against Selective forwarding and Sinkhole attacks. Authentication may be used for link layer acknowledgements also. Attacks like wormhole and hello flood cannot be countered by this approach. Insiders can still attack the network by spoofing or circulating bogus routing information. This may create sinkholes thereby allowing selective forwarding of packets. Further possibility of Sybil and hello attacks are there. For an insider using a globally shared key it may not be possible to distinguish between a genuine node or a masquerade. Following the true spirit of symmetric key cryptography each node may share a symmetric key with the base station that may be used to generate some sort of unique session key between any two neighbouring nodes with the mediation of the base station. The base station may bear the responsibility of identifying whether a particular node is asking for too many session keys and may limit such requests [47].

It is very hard to defend wormhole attack and sinkhole attacks, whether individually or combined. The use of a private channel invisible to the underlying sensor network makes detection of wormholes difficult. Routing protocols that utilise remaining energy, hop count and other information make detection of sinkholes difficult. The geographic routing protocols may be somewhat resistant to these attacks. This is because their topology is constructed on demand using local information. But here too the problem may be to identify whether location information provided by the nodes can be trusted. Considering base stations to be trustworthy, it may be assumed that it is not possible for adversaries to spoof messages from them. Additionally, authenticated broadcasts may also be used for node interactions.

As a countermeasure for selective forwarding, multipath routing seems to hold some promise. A number of disjoint routes for a source-sink pair appear to be protected against selective forwarding attack. But it may be difficult to find completely disjoint paths. Dynamic choice of the next hop may incur overhead but reduces chances of compromise.

5.6.2.1   Secure Multipath Routing Protocols

In resource-constrained wireless sensor networks, a single routing path should only be established between a source and a sink. But this implies that the path becomes non-usable once any of the intermediate nodes fails. Another event that would ultimately render the path non-usable is compromise of any node. In either situation the effect is loss of data which is unacceptable. Availability and reliability of data are necessities in any WSN application. Hence maintaining multiple paths for the same route (one source-sink pair) offers to enhance availability and reliability. However, ensuring security to multipath routing is essential for the following reasons:

- Redundant data availability – Redundant routing strategy adopted by forwarding same data packets over alternative paths makes same data available at multiple locations. The task of the adversary becomes easier.

- Route discovery attack – Attacks during route discovery will prevent the discovery of multiple available paths.

- Energy consumption – Repeated initiation of recovery mechanisms after each attack drains batteries of nodes, thereby decreasing network lifetime.

- Maintenance of multiple paths – The alternative paths must be saved in such a way that they are not compromised.

Stavrou et al. [87] presented a survey of secure multipath routing protocols for WSNs. It appears that different protocols offer different features and hence find applications for different requirements or application scenarios.

Secure and efficient intrusion-fault tolerant (SEIF) [70] is designed to be an intrusion- and fault-tolerant routing protocol for WSNs. The protocol finds node disjoint paths. The additional alternative paths start at two-hop neighbours of the sink node. Each node considers new disjoint paths coming from nodes in different sub-branches. The routing tables are maintained locally by each sensor. A node randomly chooses a path, thereby avoiding use of the same path. This protocol can fight against eavesdrop, modification, replay, Sybil and hello attacks.

Ma et al. [64] proposed a secure, multiversion, multipath (MVMP) protocol that consists of four phases. Each group of data packets is encrypted using different symmetric and asymmetric cryptographic algorithms. The encrypted packets are again reorganized into k-packet blocks and RS(n, k) coding [29] is used. Thus an n-packet RS codeword is produced for each k-packet block. Each codeword is transmitted to the destination by multiple disjoint paths. Also, packets from same codeword are transmitted via different paths. MVMP is effective against eavesdrop and modification attacks.

In INSENS [25] the authors address the DoS flooding attacks. The base station (BS) is only allowed to broadcast. It uses symmetric cryptography to provide overall security including authentication. Each node shares a secret key only with the BS. Routing tables are drawn up by the sink and provided to the nodes. The protocol constructs two alternative disjoint paths between each sensor node and sink. Each message is sent multiple times through each alternative path. The behaviour of this protocol with respect to attacks is same as that of SEIF [70], [87].

H-SPREAD [62] uses the threshold secret sharing scheme and splits a message into N pieces, called shares. Each share is forwarded by source node through different paths. The original packet is reconstructed if at least T shares are received. The multiple paths are discovered in two phases. H-SPREAD is able to address eavesdrop, modification and replay attacks.

The alternate path routing protocol SeRINS [55] proposed by Lee and Choi addresses the selective forwarding attack. It also detects and isolates malicious nodes that advertise inconsistent routing information. A neighbour node reports to the base station if it finds inconsistent routing information by a node. Then, the base station alerts the network so that sensor nodes will revoke the associated keys and exclude the malicious node. SeRINS also takes care of eavesdrop, modification, replay and sinkhole attacks.

Random network coding in directed diffusion is used by Lu et al. [63] that secures transmissions against eavesdropping attacks. Alternative routes between a source node and the sink that forward coded packets and also identify the maximum data flow that can be had per time unit.

SecMR [65] is an on-demand secure multipath routing protocol based on neighbourhood node authentication using ECC [96]. Eavesdrop, modification, replay, Sybil and hello attacks are addressed by SecMR.

The work by Song et al. [86] detects wormhole attacks in ad hoc networks as well as in WSNs. The concept of the proposed SAM protocol is based on the idea that observations and statistical analysis of information of different paths lead to the detection of the attack.

The protocol proposed by Zhao and Delgado-Frias [101] [68] is applicable for ad hoc networks including sensor networks. A feedback mechansim selects a new path when it detects nodes along a particular path dropping packets. Each source node discovers two node disjoint paths, one for data and the other for control information. Thus it works effectively against selective forwarding attacks.

Originally proposed for mobile ad hoc networks, SAODV-MAP [89] can also be applied to WSNs with appropriate asymmetric key cryptography scheme. The protocol addresses a number of attacks, like eavesdropping, modification, rushing, Sybil and hello.

Apart from the above algorithms, several other multipath routing algorithms have been proposed. These schemes are able to address different attacks, sometimes by detection only and sometimes by initiating action against the attacks. The survey by Stavrou and Pitsillides [87] presented and analysed many such algorithms.

Pre-distribution of security keys by establishment of secret key between peers is one of the major concerns in sensor networks. The multipath routing algorithms can also be used to that effect. Since compromising a path may lead to exposure of secret keys to adversaries, several methods may be adopted to further encode a key. A number of schemes like JERT [26] and a proposal by Ling and Znati [59] describe pairwise key establishment schemes. Multipath key reinforcement is also proposed by Chan et al. [19].

Multipath routing algorithms are also able to address issues like securing data aggregation in WSNs. Data integrity plays an important role for aggregator nodes to ensure that correct data is received for aggregation. Reputation-based trust management and secure multipath routing are combined in SELDA proposed by Ozdemir [71]. A sensor node computes the reputation value of its neighbourhood and determines the alternative paths. A node selects a path with higher reputation to send its data. The use of multipath routing enables the aggregator node to detect false data by comparing the multiple instances of the same data received over multiple different paths.

5.6.2.2 Energy-Efficient Secure Routing Protocols

Most routing protocols for sensor networks consider security threats and hence adopt measures to tackle them. Since consumption of energy is a major issue, many of these protocols also try to optimize energy consumption. A couple are described briefly here.

- Nidal Nasser and Yunfeng Chen propose a secure and energy-efficient routing SEER protocol [67]. The authors assume that the base station is responsible for variety of functionalities like querying data in the static network, broadcasting control packets, selective routes and maintaining the interface to outside networks. A sensor node is only responsible for functions like, sensing data, forwarding packets and sending data to base station. SEER is developed in phases beginning with topology construction. During the second stage called data transmission, the base station selects a suitable shortest path to node having data. The base station also considers energy consumption along a route while selecting the best. The last stage, route maintenance, tries to prolong network lifetime. This is done by distributing communication load among multiple routes between source and destination. Since the route is selected by the base station, the attacks on routing protocols that attract traffic by advertising high quality routes can be defended. Thus, SEER is able to defend against wormhole, sinkhole and selective forwarding attacks.

- The basic idea proposed by Sen and Ukil in [82] is single-path routing that is inherently energy-efficient. A node in a route is isolated as soon as it fails to forward packets. Packets are then routed around the node. A number of nodes in a sender's neighbourhood assume the roles of primary and secondary monitoring nodes and are responsible for forwarding packets that are dropped. This is possible because the primary and secondary monitoring nodes are aware of the sender node's other neighbours that could be utilized for the task of forwarding. To add security, packets are encrypted too. A node knows its direct neighbors by neighbor discovery and pairwise key establishment phases. This scheme is thus designed to reliably forward packets hop-wise from source to base station.

5.6.2.3   Trust-based Secure Routing Protocols

Security to routing protocols can be provided in different ways. Responsibility of finding a secure route is handled by participating sensor nodes. The nodes determine neighbours whom they can trust. The protocols described below are based on this consideration.

- A secure routing protocol that is able to adapt dynamic changes in the trust values of nodes in a network is described in [28]. A node creates a transaction of a specific number of packets when it has data to send to the base station. Each node along the path has to choose from its neighbours the best one that can deliver its data. The determination of best node is based on a metric combining the number of hops to the base station and the sender's trust of its neighbors. Each packet in the transaction uses this chosen path. After each transaction, a trust reporting phase begins, in which nodes recompute trust values for the nodes in the path of the transaction.

- Zhan and Shi consider energy of nodes while assessing trustworthiness in TARF, a trust-aware routing framework for WSNs [100]. TARF may also be coupled with existing routing protocols. Trustworthiness and energy efficiency are the main considerations of a node while choosing its neighbour from a list for forwarding data packets. The neighbours exchange energy cost report message with each other and broadcast messages from the base station are also flooded to the network periodically. The messages contain information regarding the delivery of data packets during the last interval. Such messages help a node to compute the trustworthiness of its neighbour using reports of energy cost and network loop discovery.

- Trust-based network management for hierarchical WSNs is also reported. Bao et al. proposed hierarchical trust management [7] for two-level hierarchical WSNs. Each sensor-node evaluates its neighbours in the same cluster. Similarly, each cluster-head evaluates its cluster members nodes and other cluster-heads. Both direct and/or indirect observations are taken into account for trust evaluation. Direct observations are used in case of one-hop neighbours. The nodes send their evaluations to respective cluster-heads. Each cluster-head then evaluates trust for its member nodes. Similarly, each cluster-head sends its trust evaluation results of other cluster-heads in the WSN to a cluster-head commander. This two-level trust evaluation considers four trust components: intimacy, honesty, energy and unselfishness. This trust-based management can be applied to routing and intrusion detection also.

- A lightweight and dependable trust system (LDTS) for WSNs is proposed by Li, Zhou and Du [58]. The proposal consists of a lightweight trust decision-making scheme based on the role played by a node in a clustered sensor network. A self-adaptive weighted method for trust aggregation is also proposed at the cluster-head level. This proposed methodology can be coupled with any routing scheme or platform. In this scheme both intracluster and intercluster trust evaluations consider two kinds of trust relationships: cluster member-to-cluster member direct trust and cluster head-to-cluster member feedback trust and similarly, cluster head-to-cluster head direct trust and base station-to-cluster-head feedback trust. Energy conservation is considered the basic requirement for trust calculation at the cluster-heads. This scheme is found to effectively detect and prevent malicious, selfish, and faulty cluster-heads.

5.6.2.4 Location-based Secure Routing Protocols

- One of the best known security protocols in WSN that utilize location information is secure implicit geographic forwarding (SIGF) [98]. This protocol extends the established implicit geographic routing (IGF) [9]. IGF is a stateless nondeterministic network MAC hybrid protocol. SIGF comprises three protocols: SIGF-0 is stateless and nondeterministic but also uses probabilistic candidate sampling for achieving high packet delivery ratio. SIGF-1 remembers limited (local) state and computes reputations of its neighbours (from statistics of the neighbour nodes' performances) for selecting next hop. SIGF-2 is stateful and shares state with neighbours. This may provide good defense against attack by guaranteeing authenticity, confidentiality, integrity, and freshness, but overhead is high. SIGF-0 inherits from IGF resistance to state corruption, wormhole, and hello flood attacks. SIGF-1 is able to provide resistance to a Sybil attack. SIGF-2 is able to address replay DoS attacks.

5.7 Intrusion Detection in WSN

Intrusion is an unauthorized activity in a network by means of eavesdropping (or other passive forms) or packet dropping (or other active forms). It is almost impossible to prevent intrusions in networks and intrusion detection has become the main line of defense in networks. The primary task then becomes detecting any suspicious activity in the network by its members or outsiders. Any intrusion detection system (IDS) should provide enough information for identifying the attack and its type such that proper remedial action may be taken.

5.7.1 Intrusion Detection Systems

Intrusion detection systems (IDSs) are collections of tools, methods and resources that are able to identify and report intrusions. During functioning, an IDS should not introduce too many overheads or any new vulnerability into the network. For any IDS, it is easier to find an external intruder rather than an internal intruder (a member of the network) that can become *selfish* or *malicious* after being compromised during the course of attack [14]. Intrusions may occur in various forms like disclosure of information by snooping or unauthorized access (or attempt to access), deception by masquerading, disruption in services by alteration of information, usurpation by denial of service and so on.

5.7.1.1 IDS for WSN

A large number of IDSs have already been proposed for wireless and mobile ad hoc networks. Only a few of them are applicable (with minor or major alteration) in sensor networks. The work in [14] describes several IDSs proposed only for WSNs. A brief overview on the basis of intrusion detection techniques is provided below:

- Rule-based: A set of rules is used to decide whether a given behaviour is like that of an intruder. The works in [21], [88], use rule-based detection for hierarchical sensor networks. However each of these techniques also suffers from some problems.

- Specification-based: This technique involves developing security-related specifications and observing the network for decision. The authors in [48] consider distributed WSNs and the authors in [74] consider hierarchical

WSNs but both use specification-based approaches for intrusion detection.

- Statistical-based: This technique basically detects anomalies in the network and is based on collection of data from legitimate users of the network and applying statistical tests to them. The work in [69] describes one such approach based on specific network parameters for distributed WSNs.

- Game–theoretic-based: These techniques use the concept of game theory in various forms to detect attacks and intrusions. Agah et al. [3] [2] used this approach.

- Reputation-based: Trustworthiness of nodes is considered for reputation of nodes while developing intrusion detection mechanisms. This scheme is proposed by Bao et al. [7] (trust management protocol) for hierarchical WSNs whereas Wang et al. proposed such a scheme (ranking algorithm) for sink-based IDS [92].

It is apparent from the requirement of applications that different types of IDSs are required for different applications. Resource constraints of WSN, particularly, energy, should be considered while developing IDSs. This may also influence the choice of the architecture of the sensor network.

5.8 Summary

This chapter discusses security in wireless sensor networks. It is well known that the WSN faces quite a few challenges owing to its inherent nature. The challenges stem from the sensor nodes, their deployment patterns, and data collection methodologies. Moreover, deployment of one single WSN may not serve more than one purposes. This is because the application requirements are unique in nature and generally do not provide scope for implementation in the same settings. Sometimes the network organization differs, at other times data collection methods may differ. Thus it becomes difficult to identify specific security requirements and implement them.

The chapter focuses on the different security measures that have been proposed and/or adopted so far and revealed several other issues that could not be ignored. The chapter begins with identifying security requirements of sensor networks and the attacks that a WSN may face. The attacks at different networking layers are identified. The protective measures of public key cryptography and symmetric key cryptography are discussed along with key management. Both types of WSNs are considered and a number of key generation and distribution schemes are briefly discussed.

The chapter also discusses attacks on routing protocols of WSN and their countermeasures. These include developing secure multipath routing protocols, energy-efficient secure routing protocols, trust-based routing protocols and location-based secure routing protocols. The chapter also includes discussion on intrusion detection systems of sensor networks.

This chapter also provides an overview of existing security practices.

Bibliography

[1] A. Liu and P. Ning. Tinyecc: Elliptic curve cryptography for sensor networks (version 0.1). http://discovery.csc.ncsu.edu/software/TinyECC/.

[2] A. Agah and S.K. Das. Preventing DOS attacks in wireless sensor networks: A repeated game theory approach. *International Journal of Network Security*, 5(2):145–153, 2007.

[3] A. Agah, S.K. Das, K. Basu, and M. Asadi. Intrusion detection in sensor networks: A non-cooperative game approach. In *Proceedings of 3rd IEEE International Symposium on Network Computing and Applications*. IEEE, 2004.

[4] I. F. Akyildiz and E. P. Stuntebeck. Wireless underground sensor networks: research challenges. *Ad Hoc Networks*, 4:669–686, 2006.

[5] I.F. Akyildiz, D. Pompili, and T. Melodia. Challenges for efficient communication in underwater acoustic sensor networks. *ACM Sigbed Review*, 1(2):3–8, 2004.

[6] I.F. Akyildiz, W. Su, Y. Sankarasubramaniam, and E. Cayirci. A survey on sensor networks. *IEEE Communications Magazine*, 40(8):104–112, September 2002.

[7] Fenye Bao, Ing-Ray Chen, Moon Jeong Chang, and Jin-Hee Cho. Hierarchical trust management for wireless sensor networks and its applications to trust-based routing and intrusion detection. *IEEE Transactions on Network and Service Management*, 9(12):169–183, June 2012.

[8] R. Blom. An optimal class of symmetric key generation systems. In *Proceedings of Eurocrypt*, 1984.

[9] B. Blum, T. He, S. Son, and J. Stankovic. IGF: A state-free robust communication protocol for wireless sensor networks. In *Technical Report CS-2003-11, University of Virginia, Charlottesville, VA*, 2003.

[10] C. Blundo, A. Santis, A. Herzberg, S. Kutten, U. Vaccaro, and M. Yung. Perfectly secure key distribution for dynamic conferences. In *Proceedings of Crypto*, 1992.

[11] M. Bohge and W. Trappe. An authentication framework for hierarchical ad hoc sensor networks. In *Proceedings of Workshop on Wireless Security*. ACM, 2003.

[12] D. Boneh and M. Franklin. Identity-based encryption from the Weil pairing. In *Proceedings of CRYPTO*, 2001.

[13] D. Braginsky and D. Estrin. Rumour routing algorithm for sensor networks. In *Proceedings of First ACM International Workshop on Wireless Sensor Networks and Applications*, 2002.

[14] I. Butun, S. D. Morgera, and R. Sankar. A survey of intrusion detection systems in wireless sensor networks. *IEEE Communications Surveys and Tutorials*, 16(1):266–282, 2014.

[15] S. A. Camtepe and B. Yener. Combinatorial design of key distribution mechanisms for wireless sensor networks. In *Proceedings of Ninth European Symposium on Research Computer Security*, 2004.

[16] S. A. Camtepe and B. Yener. Benchmarking block ciphers for wireless sensor networks (extended abstract). Technical Report TR–05–07, Computer Science Department, Renesselear Polytechnic Institute, 2005.

[17] D. Carman, B.Matt, and G. Cirincione. Energy-efficient and low-latency key management for sensor networks. In *Proceedings of 23rd Army Science Conference*, 2002.

[18] D. W. Carman, P. S. Kruus, and B. J. Matt. Constraints and approaches for distributed sensor network security. In *NAILabs*. Technical Report 00–010, 2000.

[19] H. Chan, A. Perrig, and D. Song. Random key predistribution schemes for sensor networks. In *IEEE Symposium on Security and Privacy*, 2003.

[20] B. Chen, K. Jamieson, H. Balakrishnan, and R. Morris. Span: an energy–efficient coordination algorithm for topology maintenance in ad hoc wireless networks. *ACM Wireless Networks Journal*, 8(5):481–494, 2002.

[21] R.C. Chen, C.F. Hsieh, and Y.F. Huang. A new method for intrusion detection on hierarchical wireless sensor networks. In *Proceedings of ICUIMC*. ACM, 2009.

[22] D. Eastlake III and P. Jones. US Secure Hash Algorithm 1 (SHA1), rfc 3174 (informational), Sept. 2001.

[23] J. Daemen and V. Rijmen. AES proposal: Rijndael. In *Proceedings of First AES Conference*. Springer, 1998.

[24] S. Datema. Case study of wireless sensor network attacks. Masters thesis, Delft University of Technology, 2005.

[25] J. Deng, R. Han, and S. Mishra. Insens: intrusion-tolerant routing in wireless sensor networks. Technical Report CUCS-939-02, Department of Computer Science, University of Colorado, 2002.

[26] J. Deng and Y.S. Han. Multipath key establishment for wireless sensor networks using just-enough redundancy transmission. *IEEE Transactions on Dependable and Secure Computing*, 5(3):177–190, 2006.

[27] W. Diffie and M. Hellman. New directions in cryptography. *IEEE Transactions on Information Theory*, 22(6):644–654, 1976.

[28] L. C. DiPippo, Y. Sun, K. Rahn Jr., R. Anachi, and O. Savas. Sarp. Technical Report, TR10–329, University of Rhode Island, Department of Computer Science, 2010.

[29] P. Djukic and S. Valaee. Minimum energy fault tolerant sensor networks. In *Proceedings of IEEE Globecom Workshops*, 2004.

[30] W. Du, J. Deng, Y. Han, S. Chen, and P. Varshney. A key management scheme for wireless sensor networks using deployment knowledge. In *Proceedings of Infocom*. IEEE, 2004.

[31] W. Du, J. Deng, Y. Han, and P. Varshney. A pairwise key pre-distribution scheme for wireless sensor networks. In *Proceedings of 10th Conference on Computer and Communications Security*. ACM, 2003.

[32] B. Dutertre, S. Cheung, and J. Levy. Lightweight key management in wireless sensor networks by leveraging initial trust. In *Technical Report SRI-SDL-04-02, System Design Laboratory, SRI International, CA*, April 2004.

[33] Elliptic Curve Cryptography. SECG Std. SEC1, 2000, www.secg.org/collateral/sec1.pdf.

[34] L. Eschenauer and V. D. Gligor. A key management scheme for distributed sensor networks. In *Proceedings of Ninth Conference on Computer and Communications Security*, 2002.

[35] A. Perrig et al. Spins: Security protocols for sensor networks. *Wireless Networks*, 8(5):521–534, 2002.

[36] T. El Gamal. A public-key cryptosystem and a signature scheme based on discrete logarithms. *IEEE Transactions on Information Theory*, 31(4):469–472, 1985.

[37] D. Ganesan, R. Govindan, S. Shenker, and D. Estrin. Highly resilient, energy-efficient multipath routing in wireless sensor networks. *Mobile Computing and Communications Review*, 4(5):11–25, 2001.

[38] P. Ganesan, R. Venugoplalan, P. Peddabachagari, A. Dean, F. Mueller, and M. Sichtiu. Analyzing and modeling encryption overhead for sensor network nodes. In *Proceedings of Second ACM International Conference on Wireless Sensor Networks and Applications*. ACM, 2003.

[39] N. Gura, A. Patel, A. Wander, H. Eberle, and S. C. Shantz. Comparing elliptic curve cryptography and RSA on 8-bit CPUS. In *Proceedings of Workshop on Cryptographic Hardware and Embedded Systems*, 2004.

[40] J. Heidemann, Y. Li, A. Syed, J. Wills, and W. Ye. Underwater sensor networking: research challenges and potential applications. In *Proceedings of WCNC*. IEEE, 2006.

[41] W. R. Heinzelman, A. Chandrakasan, and H. Balakrishnan. Energy-efficient communication protocol for wireless microsensor networks. In *Proceedings of 33rd Annual International Conference on System Sciences*, 2000.

[42] J. Hill, R. Szewczyk, A. Woo, S. Hollar, D. Culler, and K. Pister. System architecture directions for networked sensors. In *Proceedings of ACM ASPLOS IX*, 2000.

[43] Y.C. Hu, A. Perrig, and D. B. Johnson. Packet leashes: A defense against wormhole attacks in wireless networks. In *Proceedings of INFOCOM*. IEEE, 2003.

[44] J. Hwang and Y. Kim. Revisiting random key pre-distribution for sensor networks. In *Proceedings of Workshop on Security of Ad Hoc and Sensor Networks*. ACM, 2004.

[45] C. Intanagonwiwat, R. Govindan, and D. Estrin. Directed diffusion: a scalable and robust communication paradigm for sensor networks. In *Proceedings of Sixth Annual International Conference on Mobile Computing and Networks*, 2000.

[46] B. Kaliski. Twirl and rsa key size, rsa laboratories, Technical Note, may 2003.

[47] Chris Karlof and David Wagner. Secure routing in wireless sensor networks: attacks and countermeasures. *Ad Hoc Networks*, 1:293–315, 2003.

[48] I. Krontiris, T. Dimitriou, and F.C. Freiling. Towards intrusion detection in wireless sensor networks. In *Proceedings of 13th European Wireless Conference*, 2005.

[49] B. Lai and I. Verbauwhede S. Kim. Scalable session key construction protocol for wireless sensor networks. In *Proceedings of IEEE Workshop on Large Scale RealTime and Embedded Systems*. IEEE, 2002.

[50] Y. Law, R. Corin, S. Etalle, and P. Hartel. A formally verified decentralized key management for wireless sensor networks. *Personal Wireless Communications*, 2003.

[51] Y. W. Law, J. M. Doumen, and P. H. Hartel. Benchmarking block ciphers for wireless sensor networks (extended abstract). In *First IEEE International Conference on Mobile Ad Hoc and Sensor Systems*. IEEE, 2004.

[52] Y. W. Law and P. J. M. Havinga. How to secure a wireless sensor network. In *Proceedings of International Conference on Intelligent Sensors, Sensor Networks and Information Processing*. IEEE, 2005.

[53] J. Lee and D. Stinson. A combinatorial approach to key pre-distributed sensor networks. http:// www. cacr. math. uwaterloo. ca/ dstinson/ pubs.html, 2004.

[54] J. Lee and D. Stinson. Deterministic key pre-distribution schemes for distributed sensor networks. http:// www. cacr. math. uwaterloo. ca/ dstinson/ pubs.html.

[55] S. Lee and Y. Choi. A secure alternate path routing in sensor networks. *Computer Communications*, 30(1):153–165, 2006.

[56] S. B. Lee and Y. H. Choi. Arms: An authenticated routing message in sensor networks. In *Secure Mobile Adhoc Networks and Sensors, Lecture Notes in Computer Science*. Springer, 2006.

[57] M. Li and Y. Liu. Underground structure monitoring with wireless sensor networks. In *Proceedings of IPSN*, 2007.

[58] X. Li, F. Zhou, and J. Du. LDTS: A lightweight and dependable trust system for clustered wireless sensor networks. *IEEE transactions on Information Forensics and Security*, 8(6):924–935, June 2013.

[59] H. Ling and T. Znati. End-to-end pairwise key establishment using multi-path in wireless sensor network. In *Proceedings of Global Communications Conference*. IEEE, 2005.

[60] D. Liu and P. Ning. Establishing pairwise keys in distributed sensor networks. In *Proceedings of 10th Conference on Computer and Communications Security*. ACM, 2003.

[61] D. Liu and P. Ning. Location-based pairwise key establishment for static sensor networks. In *Proceedings of First Workshop on Security of Ad Hoc and Sensor Networks*. ACM, 2003.

[62] W. Lou and Y. Kwon. H-spread: a hybrid multipath scheme for secure and reliable data collection in wireless sensor networks. *IEEE Transactions on Vehicular Technology*, 55(4):1320–1330, 2006.

[63] F. Lu, L. Geng, L.T. Chia, and Y.C. Liang. Secure multi-path in sensor networks. In *Proceedings of Fifth International Conference on Embedded Networked Sensor Systems*, 2007.

[64] R. Ma, L. Xing, and H.E. Michel. A new mechanism for achieving secure and reliable data transmission in wireless sensor networks. In *Proceedings of Conference on Technologies for Homeland Security*. IEEE, 2007.

[65] M. Matsui. New block encryption algorithm misty. In *Proceedings of Fourth International Workshop on Fast Software Encryption*. Springer, 1997.

[66] A. J. Menezes, S. A. Vanstone, and P. C. V. Oorschot. *Handbook of Applied Cryptography*. CRC Press, 1996.

[67] N. Nasser and Y. Chen. Secure multipath routing protocol for wireless sensor networks. In *Proceedigns of 27th International Conference on Distributed Computing Systems Workshops*, 2007.

[68] L. B. Oliveira, H. C. Wong, M. Bern, R. Dahab, and A. A. F. Loureiro. SECLEACH: a random key distribution solution for securing clustered sensor networks. www.cs.cmu.edu/ hcwong/Pdfs/secleach.pdf.

[69] I. Onat and A. Miri. An intrusion detection system for wireless sensor networks. In *Proceedings of IEEE International Conference on Wireless and Mobile Computing, Networking and Communications*, 2005.

[70] A. Ouadjaout, Y. Challal, N. Lasla, and M. Bagaa. SEIF: secure and efficient intrusion fault-tolerant routing protocol for wireless sensor networks. In *Proceedings of Third International Conference on Availability, Reliability and Security*, 2008.

[71] S. Ozdemir. Secure and reliable data aggregation for wireless sensor networks. In *LNCS, vol. 4836*. Springer, 2007.

[72] A. S. Khan Pathan. Denial of service in wireless sensor networks–issues and challenges. *Advances in Communications and Media Research*, pages 279–309, 2010.

[73] A. Perrig, R. Canetti, J. Tygar, and D. X. Song. Efficient authentication and signing of multicast streams over lossy channels. In *Proceedings of IEEE Symposium on Security and Privacy*. IEEE, 2000.

[74] S. Rajasegarar, C. Leckie, M. Palaniswami, and J.C. Bezdek. Distributed anomaly detection in wireless sensor networks. In *Proceedings of 10th International Conference on Communication Systems*. IEEE, 2006.

[75] Recommended Elliptic Curve Domain Parameters. SECG Std. SEC2, 2000. www.secg.org/collateral/sec2.pdf.

[76] J. Rehana. Security of wireless sensor network. In *Seminar on Internetworking*. Helsinki University of Technology, 2009.

[77] R. L. Rivest. The MD5 Message-Digest Algorithm, rfc 1321, April 1992.

[78] R. L. Rivest. The rc5 encryption algorithm. *Fast Software Encryption*, pages 86–96, 1995.

[79] R. L. Rivest, M. J. B. Robshaw, R. Sidney, and Y. L. Yin. The rc6 block cipher, version 1.1, august 1998.

[80] R. L. Rivest, A. Shamir, and L. Adleman. A method for obtaining digital signatures and public-key cryptosystems. *Communication of ACM*, 21:120–126, 1978.

[81] J. Sen. Routing security issues in wireless sensor networks: Attacks and defenses. *Sustainable Wireless Sensor Networks*, pages 279–309.

[82] Jaydip Sen and Arijit Ukil. A secure routing protocol for wireless sensor networks. In *Proceedings of the ICCSA, Part III, LNCS 6018*. Springer, 2010.

[83] A. Shamir. Identity-based cryptosystems and signature schemes. In *Proceedings of CRYPTO*, 1984.

[84] E. Shi and A. Perrig. Designing secure sensor networks. *Wireless Communication Magazine*, 11(6):38–43, 2004.

[85] S. Slijepcevic, M. Potkonjak, V. Tsiatsis, S. Zimbeck, and M. B. Srivastava. On communication security in wireless ad hoc sensor networks. In *Proceedings of 11th International Workshop on Enabling Technologies: Infrastructure for Collaborative Enterprises*. IEEE, 2002.

[86] N. Song, L. Qian, and X. Li. Wormhole attacks detection in wireless ad hoc networks: a statistical analysis approach. In *Proceedings of 19th International Parallel and Distributed Processing Symposium*. IEEE, 2005.

[87] E. Stavrou and A. Pitsillides. A survey on secure multipath routing protocols in WSNS. *Computer Networks*, 54:2215–2238, 2010.

[88] C.C. Su, K.M. Chang, Y.H. Kuo, and M.F. Horng. The new intrusion prevention and detection approaches for clustering-based sensor networks. In *Proceedings of Wireless Communications and Networking Conference*. IEEE, 2005.

[89] B. Vaidya, J.Y. Pyun, J.A. Park, and S.J. Han. Secure multipath routing scheme for mobile ad hoc network. In *Proceedings of Third IEEE International Symposium on Dependable, Autonomic and Secure Computing.* IEEE, 2007.

[90] John Paul Walters, Zhengqiang Liang, Weisong Shi, and Vipin Chaudhary. Wireless sensor network security: A survey. *Security in Distributed, Grid, and Pervasive Computing,* pages 1–50.

[91] A. S. Wander, N. Gura, H. Eberle, V. Gupta, and S. C. Shantz. Energy analysis of public-key cryptography for wireless sensor networks. In *Proceedings of Third International Conference on Pervasive Computing and Communication.* IEEE, 2005.

[92] C. Wang, T. Feng, J. Kim, G. Wang, and W. Zhang. Catching packet droppers and modifiers in wireless sensor networks. *IEEE Transactions on Parallel Distributed System,* 23(5):835–843, 2012.

[93] X. Wang, L. Yang, and K. Chen. SDD: Secure directed diffusion protocol for sensor networks. In *Proceedings of ESAS, LNCS 3313.* Springer, 2004.

[94] Yong Wang, Garhan Attebury, and Byrav Rammurthy. A survey of security issues in wireless sensor networks. *IEEE Communications Surveys and Tutorials,* 2nd Quarter:1–23, 2006.

[95] R. Watro, D. Kong, S. Cuti, C. Gardiner, C. Lynn, and P. Kruus. Tinypk: Securing sensor networks with public key technology. In *Proceedings of Second Workshop on Security of Ad Hoc and Sensor Networks.* ACM, 2005.

[96] A. Weimerskirch, C. Paar, and S. C. Shantz. Elliptic curve cryptography on a palm OS device. In *Proceedings of Sixth Australasian Conference on Information Security and Privacy.* Springer, 2001.

[97] D. J. Wheeler and R. M. Needham. TEA, a tiny encryption algorithm. In *Proceedings of Second International Workshop on Fast Software Encryption, Vol. 1008.* Springer, 1994.

[98] A. D. Wood, L. Fang, J. A. Stankovic, and Tian He. SIGF: a family of configurable, secure routing protocols for wireless sensor networks. In *Proceedings of Fourth ACM Workshop on Security of Ad Hoc and Sensor Networks.* ACM, 2006.

[99] Y. Xu, J. Heidemann, and D. Estrin. Geography-informed energy conservation for ad hoc routing. In *Proceedings of Seventh Annual ACM/IEEE International Conference on Mobile Computing and Networking,* 2001.

[100] G. Zhan, W. Shi, and J. Deng. Design and implementation of TARF: A trust-aware routing framework for WSNS. *IEEE Transactions on Dependable and Secure Computing*, 9(2):184–197, April 2012.

[101] L. Zhao and J.G. Delgado-Frias. Multipath routing-based secure data transmission in ad hoc networks. In *Proceedings of International Conference on Wireless and Mobile Computing, Networking and Communications*. IEEE, 2006.

[102] S. Zhu, S. Setia, and S. Jajodia. LEAP: Efficient security mechanisms for large-scale distributed sensor networks. In *Proceedings of 10th Conference on Computer and Communications Security*. ACM, 2003.

6

Operating Systems for WSNs

6.1 Introduction

In the recent years, there have been several research efforts to develop operating systems (OSs) for sensor networks. The role of any OS is to support development of reliable application softwares by providing a convenient and safe abstraction of hardware resources. The hardware design of a sensor network is driven by its application-specific requirements. Requirements from processing capabilities to radio bandwidth and sensor modules led to a huge variety of hardware components. That is the reason for making sensor networks hardware modular and heterogeneous. In general, a wireless sensor network (WSN) operates both at network level and node level. Network level requirements are mainly connectivity and communication. On the other hand, nodes have constraints of available resources like processing power, memory, radio frequency and limited battery power. Thus, OS support is crucial for sensor network applications to ensure connectivity and hardware abstraction along with management of limited resources. The basic functionality of an OS for a WSN is to hide the low-level details of the sensing device by providing an interface to the application layer.

In conventional computing, the OS allocates application processing threads to processors, maps virtual addresses to locations in memory, and manages disks, networks, and peripherals on the application's behalf. In WSN systems, sensors can be located far away from the networked infrastructure and easy human accessibility. Wireless sensor nodes manifest characteristics of both embedded systems and general purpose systems. They must use less energy and be robust to environmental conditions while providing common services that make it easy to write applications. The design of an OS for a WSN deviates from traditional operating system design due to significant and specific characteristics like constrained resources, high dynamics and inaccessible deployment. In general, operating systems for sensor networks require a complete binary image of the entire system built and downloaded into each device. The binary image includes the operating system, system libraries, and the actual applications running on top of the system. Architecture, execution model, reprogramming, scheduling, power management and simulation environment are the important OS features that classify the existing WSN operating systems.

The classification helps clarify the contrasting differences in existing operating systems for sensors.

6.2   Architecture

The architecture of an OS influences: (a) run-time reconfigurability of the services, and (b) size of the core kernel. The addition of new kernel services or updating them depends entirely on the architecture of the operating system. The architecture must allow extensions of the kernel if required. Also the architecture must be flexible i.e., only application-required services may be loaded onto the system. Size of the core kernel is very important due to the strict resource constraints on the sensor nodes. The limited memory of only a few kilobytes on a sensor node necessitates the OS to be designed with a very small footprint. There are mainly three kinds of architectures cited in the literature: (1) monolithic (2) micro-kernel and (3) virtual machine.

6.2.1   Monolithic

The monolithic approach of building a kernel is straightforward and does not follow any specific structure. A monolithic kernel is an operating system architecture where the entire operating system is working in the kernel space. All functionality provided by the operating system is realized within the kernel. The monolithic architecture defines a high-level virtual interface over hardware, with a set of primitives or system calls to implement all operating system services. Services provided by an operating system are implemented separately and each service provides an interface for other services. The kernel consists of a set of procedures which are able to call each other without any restrictions. This is represented in Figure 6.1. One advantage of the monolithic architecture is that the module interaction costs are low. All the services that constitute core kernel may not be required all the time for the running applications. WSN operating systems are required to have extremely small memory footprints. The monolithic architecture allows a tight integration of required operating system components together into a single system image. However, such an architecture can also support building an application-specific single-system image kernel that binds only the required services for an application. Although this reduces the size of the kernel, it does not allow running of multiple applications. Moreover, such an architecture requires an entire image to be replaced if there are any changes to the kernel or application. In such architecture, it is very difficult to maintain a system that becomes hard to understand and modify. Popular operating systems for sensors which fall partly under monolithic architecture are TinyOS and MagnetOS.

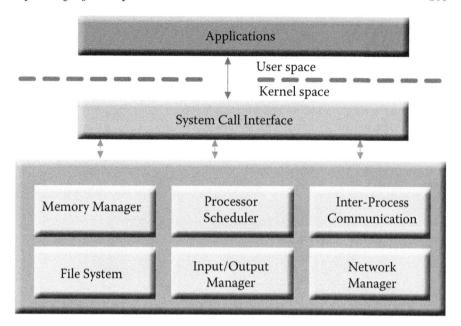

FIGURE 6.1
Monolithic OS [16]

6.2.2 Micro-kernel

A micro-kernel OS is structured as a tiny kernel that provides the essential services used by a set of processes, which in turn provide the higher level OS functionality. The real goal in designing a micro-kernel OS is not simply to make it small. A micro-kernel OS embodies a fundamental change in the approach to deliver OS functionality. Modularity is the key to the design. Micro-kernel operating systems provide a more structured architecture than conventional, monolithic kernels. The micro-kernel approach is to define a very simple abstraction over the hardware with a set of primitives or system calls to implement minimal OS services such as thread management, address spaces and interprocess communication. Figure 6.2 shows a model micro-kernel approach. All other services, those normally provided by the kernel such as networking, are implemented in user-space programs referred to as servers. The main design issue in micro-kernel architectures is the definition of the services needed at the micro-kernel interface. This interface must represent a good balance between fully supporting functions that are key to all systems, enabling customization to adapt the system to the requirements of specific application areas and taking advantage of particular machine architectures. A generic micro-kernel provides support for basic operations such as processor real-time scheduling, (virtual) memory management and local-

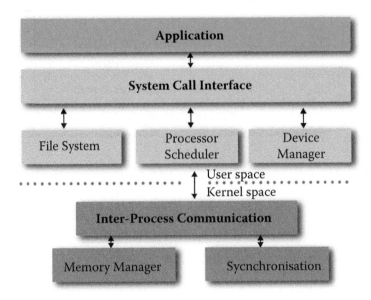

FIGURE 6.2
Micro-kernel OS [16]

ization transparent interprocess communication facility between servers that
implement more complex operating system-dependent functions. Micro-kernel
based architecture is often justified as an approach to reducing the size and
complexity of the OS. Restructuring the OS with a better architecture reduces
the quantity of code required.

6.2.3   Virtual Machine

Generally, a virtual machine (VM) can circumvent real machine compatibility
constraints and hardware resource constraints to enable a higher degree of
software portability and flexibility. The software running on a VM is tailored
to the resources and abstractions provided by the VM. A system VM provides
a platform for the execution of a complete OS and example are VMWare and
Xen. On the other hand, a process VM provides a platform for the execution
of a single program or process. Examples are Java VM, .NET VM. An OS
that executes in the context of a virtual machine is often referred to as a
guest OS and the VM may sometimes be called a "hypervisor." A hypervisor
can run on bare hardware (native VM) or on top of an OS (hosted VM).
The VM is implemented as a combination of a real machine and virtualizing
software. The hypervisor provides access to the hardware resources for the
operating systems via fully or partially virtualized interfaces. The VM bears
almost no functional resemblance to the underlying hardware which actually

performs the work. In general, a virtual operating system is restricted to parts of an ordinary OS which are important in completing the designated tasks. The simplest implementation of such an OS is to simulate the virtual machine's instruction set in software. However, efficient implementations use emulation. This requires that the underlying real hardware has mostly the same instruction set as the VM. Architecture of a VM-based OS is shown in Figure 6.3.

Instead of a common VM as the node runtime, like a Java virtual machine (JVM), a sensor network uses an application-specific VM. Virtualizing application code separates programs from the underlying hardware and operating system. A virtual architecture can easily provide alternative execution and data models such as lightweight threads or declarative queries. Virtualization also has the traditional benefit of platform independent code. Different hardware platforms can implement the same virtual machine, allowing a user to reprogram a heterogeneous network uniformly. As sensor network deployments are application domain-specific, a virtual machine can optimize its instruction set with domain-specific information.

6.3 Execution Model

6.3.1 Event-based OS

An event-driven OS is more like a system in which conceptually concurrent components are activated or change state only in response to an incoming event. A simple event-driven execution model is shown in Figure 6.4. An event is an independent action which occurs in the system and may interact with other events. Events are signal-driven and can be triggered by a function which matches a particular event. An event is activated only after receiving a trigger signal from an interrupt service routine (ISR). The event handler carries out the action. In essence, the event-based model represents an approach for designing loosely coupled systems with expressive interaction mechanisms. Sometimes it is assumed that an event-driven OS is more suitable for sensor networks because fewer resources are needed, resulting in a more energy-efficient system.

The OS used for the sensor nodes influences energy consumption on two levels. First, the design of the OS defines the minimum resource requirements such as CPU speed and memory capacity that the sensor hardware must provide. Second, the OS design also impacts the usage pattern of the CPU and thus determines how often energy-efficient sleep periods can be activated. Event-driven execution is particularly suitable for untethered devices such as sensor nodes. The sensor can be put into sleep mode to preserve energy when no interesting events are happening. To ensure long periods of unattended

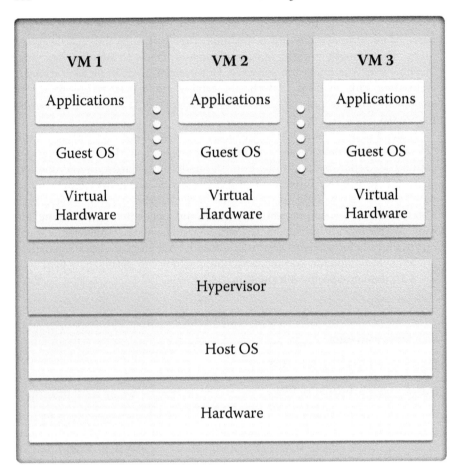

FIGURE 6.3
Virtual machine-based OS (Source: http://sp.parallels.com/products.pcs/
hypervisor/)

network operation, the energy consumption of the sensor nodes must be very
low.

6.3.2 Thread-based OS

In a thread based OS, there is at least one thread for every process, which
is called the primary thread. This thread is created as soon as its mother
process is created. This methodology increases OS efficiency through better
utilization of main memory and reduction of competition for resources. In
this model, kernel activities are performed by a set of threads. As a result,

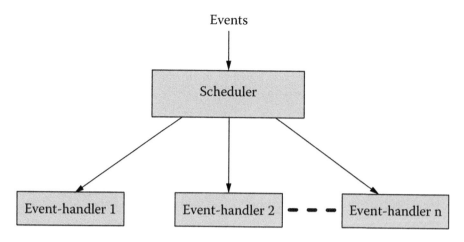

FIGURE 6.4
A simple event-driven execution model

code reusability is achieved and kernel size is reduced. Figure 6.5 shows a thread-driven execution model.

Threads are like functions, each with its own stack and thread control block (TCB). Some threads have real-time requirements. They have to respond quickly and correctly. The term multithreading is also used to describe the situation of allowing multiple threads in the same process. When a multithreaded process is run on a single-CPU system, threads are basically processes that run in the same memory context. When a thread is finished running, it can save the information in the thread table and then call the thread scheduler to pick another thread to run. No kernel call is required for this. Moreover, the procedure that saves the thread's state and the thread scheduler are just the local procedures. So, invoking them is much more efficient than making a kernel call. No trap is needed, no context switching occurs, and memory need not be flashed. This makes thread scheduling very fast. A thread-driven approach is attractive in sensor networks for above-mentioned features.

6.3.3 Hybrid Models

Both event-based and thread-based approaches have their pros and cons. Event-based designs are not suitable for managing concurrency. Control flow in event-based systems is difficult to achieve. On the other hand, threads are not always effective due to the difficulties of ensuring proper synchronization. Although threads can be made lightweight and efficient, an event-based system still has the advantage in terms of flexibility and customizability. Usually, the programmer would design parts of the system using threads where threads

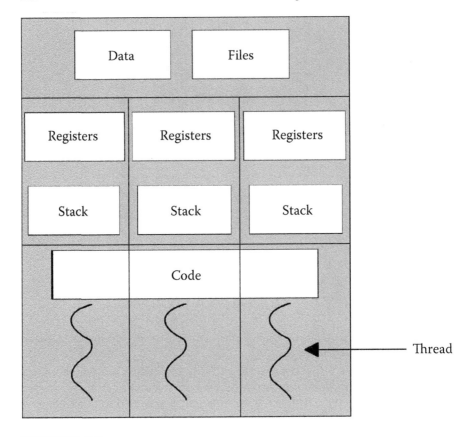

FIGURE 6.5
A simple thread-driven execution model

are the appropriate abstractions and parts of the system using events where they are more suitable. Many existing systems implement the hybrid model to various degrees. In general, they have a bias either toward events or threads.

To make the hybrid model feasible, the thread scheduler interface provides a certain level of abstraction in the form of event handlers. Every thread is assigned a priority and the thread with the highest priority is executed. The order of execution depends on this priority. The rule is very simple. The scheduler activates the thread that has the highest priority of all threads that are ready to run. The scheduler is a customizable event-driven system. Figure 6.6 presents the hybrid model that can be seen either as a multithreaded system with a scheduler or as an event-driven system with a thread abstraction for representing control flow. The main objective is to implement both thread abstractions and event abstractions. Thread abstractions represent control flow whereas event abstractions provide scheduling primitives.

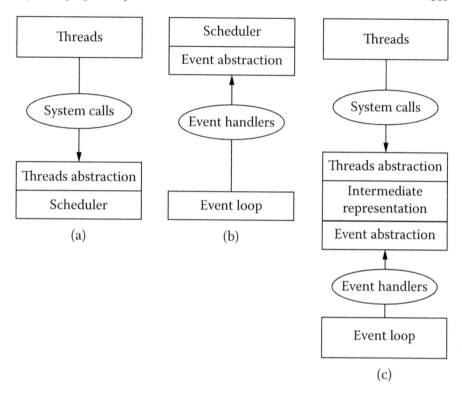

FIGURE 6.6

(a) Thread model, (b) event model, (c) hybrid model

6.4 Scheduling

Scheduling involves determining the order of execution for a set of tasks with certain known characteristics of a limited set of resources. In WSNs there are different resources like the CPU, memory, transmitter, receiver and battery for which scheduling techniques are required.

There are two kinds of constraints faced by tasks executing in sensor network environment: *time critical* and *resource critical*. The selection of an appropriate scheduling algorithm for WSNs typically depends on the nature of the application. Real-time scheduling algorithms are suitable for applications having real-time requirements and thus applicable to preemptive operations. For other applications, non-real-time scheduling algorithms are sufficient and also applicable to non-preemptive functions. Event handlers can preempt the current running task and this can affect time-critical application. The other issue that affects the design of a WSN system is resource constraint. WSNs face limited RAM availability, limited processing power, energy consumption constraint, channel access and transmission and reception power constraints. A WSN system is designed to work optimally in spite of its resource constraint problems.

Scheduling policies for real time systems are classified as:

- *Static or pre-runtime*: Tasks execute in a statically decided order. The order of execution is decided before the tasks are entered into the system based on statically defined criteria, e.g., periodicity. The advantage of static scheduling is that it involves almost no overhead in deciding which task to schedule. However, the lack of flexibility makes it difficult to model the system accurately and provide correct optimizations. This causes systems to be over-constrained due to statically decided scheduling.

- *Dynamic or runtime*: The order of execution is decided on the basis of priority. The tasks are dynamically chosen from an ordered prioritized queue. The priorities can be assigned based on different criteria, e.g., deadline, criticality, periodicity and other factors. Dynamic scheduling can be preemptive or non preemptive.

WSNs are used in both real-time and non-real-time environments. A WSN OS must provide scheduling algorithms that can accommodate the application requirements. Moreover, a suitable scheduling algorithm should be memory- and energy-efficient. Thus a hybrid approach can also be used where tasks are divided and placed into different queues at different scheduling levels. A scheduler could consist of two levels of scheduler queues, the upper queue consisting of very high priority (critical) non-preemptive tasks and a lower queue for preemptive tasks.

6.5   Power Management

The key concern in WSNs is energy consumption, as sensor nodes are battery powered. The battery has limited capacity and often can neither be replaced nor recharged due to environmental or cost constraints. Power management is very essential in the deployment of WSNs in a variety of applications including habitat monitoring, structural monitoring and surveillance. WSNs in these applications must remain operational for long lifetimes on limited energy resources. For example, due to high cost of embedding sensors in buildings or bridges, a WSN deployed for structural health monitoring should be autonomous and self-sustainable, able to function for several years with a battery power supply to be economically viable. Although energy harvesting techniques (e.g., solar, vibrational or wind energy) can provide additional energy, the amount of energy available remains scarce.

Usually, sensing applications present a wide range of requirements in terms of data rates, computation and communication distance. MAC and routing protocols and algorithms have to be modified for each application. A sensor node's lifetime is defined as the node's operating time without the need for any external intervention like battery replacement. A WSN lifetime can be defined as the lifetime of the shortest living node in the network (critical node). However, depending on the application, density of the network and possibilities of reconfiguration, it can be defined as the lifetime of some other (main or critical) node. Efficient power management in the sensor nodes is a determining factor for the network's lifetime. In order to prolong a WSN lifetime, it is required to reduce the energy consumption of the nodes as much as possible and form an energy-aware system. Typically the wireless radio of a sensor node consumes the highest energy share (much more than the sensing and processing components). For example, in a typical sensor node, the energy cost of transmitting a bit of information is approximately the same as the cost of executing a thousand operations [22].

There are five modes of sensor nodes as shown in Figure 6.7 (e.g., TelosB mote with radio CC2420) [9] with respect to operations: sense mode, computation, transmission mode, receiving mode, idle mode and sleep mode. Table 6.1 shows energy consumption of different components in TelosB [10]. Furthermore, the power consumed when the radio transceiver is idle is often nearly the same as the power consumed in the transmit or receive mode. Many traditional MAC protocols used in wireless sensor networks, such as 802.11, require the sensor to remain awake to listen the medium, even when the sensor is not transmitting or receiving a packet. This idle listening mechanism is inefficient and wastes significant energy if there is light traffic on the network. To mitigate this energy consumption of idle listening, duty cycling mechanisms have been introduced in sensor network MAC protocols.

TABLE 6.1

Energy consumption of different components in TelosB [10]

Module	Power	Mode
Processor/memory	$1.8mA$	Active
Processor/memory	$5.1\mu A$	Sleep
Radio Rx mode	$18.8mA$	Receiving
Radio Tx mode	$17.4mA$	Transmission
Radio Idle mode	$21\mu A$	
Radio Sleep mode	$1\mu A$	

Thus the most effective way to reduce energy consumption is to switch off the transceiver when communication is not needed. In this situation sensor nodes alternate between sleep and wakeup periods, and neighboring nodes coordinate themselves by implementing a sleep and wakeup schedule in order to make communication successful. This technique is referred as duty cycling. A runtime support system for sensor networks applications should provide

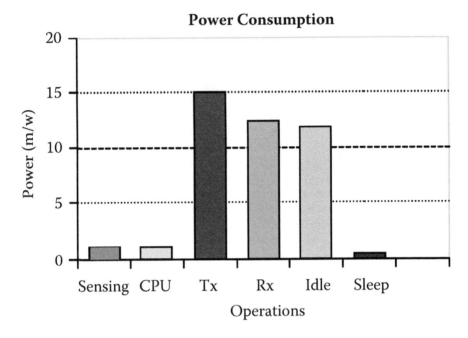

FIGURE 6.7 (SEE COLOR INSERT)
Energy consumption at different states

power management mechanisms to the applications, and use as little power as possible to provide its services. Thus power management in operating systems and software is critical for WSNs because programmability is a requirement. The operating system manages the resources on each node, provides a layer of abstraction for the hardware, and gives the system developer a programming interface that allows applications to be efficiently implemented.

6.6   Communication

Sensor networks are inherently communication-centric systems. In general, sensor readings can be transported directly from a source sensor node (single-hop) or through neighbor nodes (multi-hop) to the base station or sink. Therefore, efficient network protocols are required to build a network of sensors. The network protocols define the message formats and sequences of steps for the nodes to follow to communicate with each other. For network protocols to operate, an OS that implements these protocols runs on every node. A protocol stack consists of a number of layers in a standardized order. Each layer is designed with a single purpose, which in combination with other layers allows applications on different nodes to exchange data over a network. Given the specific communication requirements of different applications, communication hardware for sensor networks should be widely configurable. The OS should provide means to configure the communication protocol stack. Sensor network communication is made challenging by the physical communication environment and resource constraints of the individual nodes. Thus, OS must effectively manage the physical communication medium as well as constrained resources and should provide an efficient programming interface which allows programmers to create resource-efficient applications.

A communication protocol stack involves application, transport, routing, medium access control (MAC) and physical layers (see Figure 2.1 in Chapter 2). It enables the information exchange among the nodes of the network and produces a change in the state of the network. The design of a cross-layer optimization algorithm for WSNs that considers both performance and energy factors requires efficient communication between protocol stack layers. Generally, the communication protocols deal with the construction of physical channels, the access of the node into the medium (MAC), the selection of routing path through which the node transmits its data and the transport of packets from one node to another node.

The *physical layer* address the needs of reception and transmission techniques. Frequency selection, carrier frequency generation, signal detection, modulation and signal-to-noise ratio are handled in this layer.

The *data link layer* is responsible for the multiplexing of data streams, data frame detection, medium access and error control. Control access to the shared medium is handled in this layer. It ensures reliable point-to-point and point-to-multipoint connections in a WSN environment.

The *network layer* is responsible for multi-hop wireless routing protocols between source sensor nodes and the sink node. Energy-efficient routes, data aggregation, attribute-based addressing and location awareness are handled in this layer.

The *transport layer* is responsible for efficient and secure send and receive messaging over a WSN and also controls congestion. This layer also plays an important role when the WSN is planned to be accessed through the Internet or other external networks.

The *application layer* is closely related to the specific application and environment and its design depends on the specific application demand. The main task of this layer is the acquisition of data.

The functions performed by the management layer can be categorized into the following three areas:

- *Power management* is responsible for energy usage of a node. For example, when the power level of a sensor node is low, this layer may decide to restrict participation of that node in routing messages.

- *Mobility management* supports mobility-aware protocols. For example, it

detects and registers the movements of nodes so that network connectivity
is always maintained.

- *Task management* is responsible for scheduling. For example, all sensors
  may not required to perform a sensing task and some sensors may be
  selected on the basis of their power levels to perform the task.

In this way, sensors work together to provide energy-efficient, mobility-aware
schemes. Collaboration among the sensors in routing and data collection protocols also prolongs the network lifetime.

6.7   Case Study: Popular Operating Systems

In the following subsections, we describe state-of-the-art operating systems
for sensor networks.

6.7.1   TinyOS

TinyOS [15], developed at the University of California at Berkeley, is possibly the earliest operating system for sensor networks [18]. The main goals of
TinyOS [19] are summarized as:

- To consider present and future designs for sensor networks and sensor
  network nodes.

- To support diverse implementations of operating system services and applications targeting different hardware and software mixes.

- To address the challenges of sensor networks: limited resources,
  concurrency-intensive operation, application-specific requirements and a
  need for robustness.

TinyOS provides an efficient framework for modularity and a resource-
constrained, event-driven concurrency model to achieve these goals. The modularity framework of TinyOS allows the system to adapt to hardware diversity
and allows applications to reuse common software services and abstractions.

The design of TinyOS is based on the event execution model. Design patterns used by software components in TinyOS differ significantly from traditional software design patterns due to the constraints of sensor networks.
Generally, design patterns identify common and recurring requirements and
define patterns of object interactions that meet these requirements. Even so,
these patterns are not directly applicable to TinyOS programming. The main
focus of TinyOS design is on static allocation and whole-program composi-
tion. This helps provide a high modularity and offers a constrained runtime

environment for sensors. Networks of sensors require concurrency-intensive operations and are constrained by minimal hardware resources. TinyOS is a set of components included as needed in applications. However, one of the significant challenges in TinyOS is the creation of flexible, reusable components. The component-based runtime environment of TinyOS provides support to WSN systems. This runtime environment is developed using network embedded system C (nesC) language [11], a high-level language for building structured component-based applications. nesC is a C dialect with some additional language features for components and concurrency. It is a component-based language with an event-based execution model. nesC has some similarities and dissimilarities with object-oriented languages. In nesC, components can encapsulate state and interact through well-defined interfaces similar to objects. On the other hand, the set of components and their interactions are fixed at compile time to provide reliability and efficiency, rather than at runtime, as the case in object-oriented references and instantiation.

TinyOS provides a set of important services and abstractions, such as sensing, communication, storage, and timing. Three essential differences between C and nesC are components, interfaces and wiring all relate to naming and organizing a program's elements (variables, functions, types, etc.). At a high level, TinyOS provides three services to develop an application:

- Component model defines how to write small, reusable pieces of code and compose them into larger abstractions.

- Concurrent execution model defines how components interleave their computations and interrupt and non-interrupt code interactions.

- Application programming interfaces (APIs) define services, component libraries and an overall component structure that simplify writing new applications and services.

6.7.1.1 Components and Interfaces

Programming model of TinyOS centers around components that encapsulate a specific set of services which are specified by interfaces. A nesC application consists of one or more components assembled or wired to form an executable application. TinyOS provides a large number of components along with a task scheduler to application developers, including abstractions for sensors, single-hop networking, ad hoc routing, power management, timers, and non-volatile storage. Components define two scopes: one for their specification which contains the names of their interfaces, and a second scope for their implementation. An application connects components using a *wiring specification* that is independent of component implementations. The complete set of components that the application uses is defined by the wiring specification. A component has two classes of interfaces: those it *provides* and those it *uses*.

TABLE 6.2

Core interfaces provided by TinyOS [20]

Interface	Description
Clock	Hardware clock
EEPROMRead/Write	EEPROM read and write
HardwareId	Hardware ID access
I2C	Interface to I2C bus
Leds	Red, yellow and green LEDs
MAC	Radio MAC layer
Mic	Microphone interface
Pot	Hardware potentiometer for transmit power
Random	Random number generator
ReceiveMsg	Receive active message
SendMsg	Send active message
StdControl	Initiate start and stop components
Time	Get current time
TinySec	Lightweight encryption and decryption
WatchDog	Watchdog timer control

- The *provided interfaces* are intended to represent the functionality that the component provides to its user in its specification.

- The *used interfaces* represent the functionality the component needs to perform its job.

Table 6.2 lists core interfaces that are available to application developers [19].

In general, a developer composes an application by writing components and wiring them to TinyOS components that provide implementations of the required services. A component can provide or use the same interface type several times as long as it gives each instance a separate name. Basically, these interfaces define how the components directly interact with each other. An interface generally models some service (e.g., sending a message, receiving a message) and is specified by an interface type. The state machine for the SendMsg interface is shown in Figure 6.8.

As shown in Figure 6.9, the TimerM component which is a part of the TinyOS timer service provides the StdControl and Timer interfaces and uses a Clock interface. Figure 6.10 shows some sample TinyOS interface types, for example, the basic packet communication interface, SendMsg. Rather than wait until an operation (e.g., SendMsg.send) completes, the interface command returns immediately, allowing the application to continue processing. When the operation does complete, the interface signals the completion event (e.g., SendMsg.sendDone), at which point the user can reclaim the packet buffer.

Components have three computational abstractions: *commands*, *events* and *tasks*. Commands and events are mechanisms for inter-component communication and tasks are used to express intra-component concurrency.

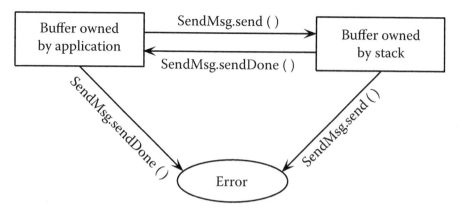

FIGURE 6.8
State machine for SendMsg interface [19]

- A *command* is a request to a component to perform some service, for example, initiate a sensor reading. Commands are non-blocking and return immediately. Typically, a command will deposit request parameters into its frame and conditionally post a task for later execution. It may also invoke commands on lower components. A command must also provide feedback to its caller by returning status indicating whether the service was successful or not.

- An *event* is used to signal the completion of a service initiated by a command. Events such as message arrival may also be signaled asynchronously. Events are also non-blocking and signal completion at a later time.

```
module TimerM {
    provides {
        interface StdControl;
        interface Timer{uint8_t id];
    }
    uses interface Clock;
}
implementation {
    ... a dialect of C ...
}
```

FIGURE 6.9
TimerM component [19]

```
/* interface of StdControl for commands */
interface StdControl {
    command result_t init ();
    command result_t start ();
    command result_t stop ();
}
/* interface of Timer for commands and events*/
interface Timer {
        command result_t start (char type, unit32_t interval);
        command result_t stop ();
        event result_t fired ();
}
/* interface of Clock for commands and events*/
interface Clock {
        command result_t setRate (char interval, char scale);
        event result_t fire ();
}
/* interface of SendMsg for commands and events*/
interface SendMsg{
        command result_t send (unit16_t address units8_t length,
        TOS_MsgPtr msg);
        event result_t sendDone (TOS_MsgPtr msg,result _t success);
}
```

FIGURE 6.10
Some TinyOS interface types

- A *task* is posted by commands and event handlers. A task is a function whose execution is deferred. Rather than performing a computation immediately, a task is executed by the TinyOS scheduler at a later time. This is the reason for commands and events to be responsive and non-blocking. Tasks represent internal concurrency within a component and may only access state within that component. The basic execution model of a task is run-to-completion, rather than to run-indefinitely. Thus, tasks are much lighterweight than threads. As tasks run to completion and do not preempt each other, there is no need to manage context switching among them.

The tasks are implemented inside the application. As shown in Figure 6.11, the following section of source code implements the action of switching the red LED on and off for every **Timer** event. For that, it defines a task (processing) that implements the action. The task post is done at the timer event handler (**Timer.fired**):

Interfaces are bidirectional: they specify a set of commands, which are functions to be implemented by the interface's provider, and a set of events, which are functions to be implemented by the interface's user. For a component to call the commands in an interface, it must implement the events that interface. A single component may use or provide multiple interfaces and multiple instances of the same interface. The set of interfaces which a component provides together with the set of interfaces that a component uses constitute

```
/* The task is triggered by the Timer.fired event which toggles the
red LED */
task void processing ()
{
        if(state)
          call Leds.redOn();
        else
          call Leds.redoff();
}
/* Using the processing task, toggle the red LED in response to the
Timer.fired event */
event result_t Timer.fired()
}
        state= !state;
        post processing();
        return SUCCESS;
}
```

FIGURE 6.11
Example of source code that implements a task post

that component's signature. Two types of components are available in nesC: *modules* and *configurations*.

- *Modules* provide code with extensions for calling and implementing commands and events. A module declares private state variables and data that only it can reference. Modules are written in a dialect of C.

- *Configurations* are used to wire other components together. They connect interfaces used by one components to interfaces provided by other components. Configurations allow multiple components to be aggregated together into a single super-component that exposes a single set of interfaces.

Modules and configurations have names specifications and implementations. Each component has its own interface namespace, which it uses to refer to the commands and events. A configuration wiring interfaces by making the connection between the local name of an interface used by one component to the local name of the interface provided by another. An implementation cannot name another component. That means components interact solely through interfaces and invoke an interface without referring explicitly to its implementation. Figure 6.12 describes the TinyOS timer service, which is a configuration (TimerC) that wires the timer module (TimerM) to the hardware clock component (HWClock).

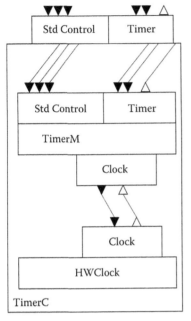

```
configuration TimerC {
      provides {
            interface StdControl;
            interface TImer[uint8_t id];
      }
}
implementation {
      components TimerM, HWClock;

      StdControl = TimerM.StdControl;
      Timer = TimerM.Timer;

      TimerM.Clk → HWClock.Clock;
}
```

FIGURE 6.12
The `TimerC` configuration [20]

6.7.1.2   Concurrent Execution Model

Concurrency control plays a significant role in the event-centric domain of sensor networks. Rather than performing computation-intensive tasks, sensor nodes predominantly process multiple information flows "on the fly." Nodes must simultaneously execute several operations but have limited storage. For example, events can arrive at any time and must interact with the ongoing computation.

In this situation generally two approaches are used: (a) run later (b) run immediately. TinyOS implements the latter approach. The concurrent execution model enables TinyOS to support many components needing to act at the same time while requiring little RAM. First, every I/O call in TinyOS is split-phase: rather than block until completion, a request returns immediately and the caller gets a callback when the I/O completes. Since the stack is not tied up waiting for I/O calls to complete, TinyOS only needs one stack and does not have threads.

In TinyOS, the core of the concurrent execution model is composed of *tasks* and *hardware event handlers*. As discussed earlier, tasks are functions whose executions are deferred. Tasks are explicit entities implemented inside an application. A program submits a task to the TinyOS scheduler for execution.

Tasks are posted to a queue. As tasks are processed, interrupts can trigger hardware events that preempt tasks. When the task queue is empty, the system goes into a sleep state until the next interrupt. If this interrupt queues a task, TinyOS pulls it off the queue and runs it. If not, the system returns to sleep. The scheduler can execute tasks in any order. The standard TinyOS scheduler follows a FIFO[1] policy. Another policy called earliest deadline first is also implemented in TinyOS. A task does not have a private context, that is, all the tasks share the same context of execution. As tasks run to completion and there is no preemption between tasks, there is no need of storing context information in each task. Thus, tasks are atomic with respect to each other. However, tasks are not atomic with respect to interrupt handlers or to commands and events they invoke. Hardware event handlers are executed in response to a hardware interrupt. Hardware events are interrupts caused by a timer, sensor, or communication device. They run to completion but may preempt the execution of a task or other hardware event handler. The context switching between an activated hardware event handler and a task or other hardware event handler is done automatically without the need of any special context management. The interrupt routine handler simply saves the status when the preemption begins and restores it when the task ends.

TinyOS differentiates synchronous and asynchronous code to facilitate the detection of race conditions.

- *Synchronous code* is only reachable from tasks.

- *Asynchronous code* is reachable from at least one interrupt handler.

Components often have a mix of synchronous and asynchronous codes. Here, the main objective is to allow developers to build responsive concurrent data structures that can safely share data between synchronous and asynchronous codes. Even though there are no races among tasks due to their non-preempt nature, there are still potential races between synchronous and asynchronous codes and sometimes between asynchronous codes. In general, any update to shared state that is reachable from asynchronous code is a potential data race. To address this issue, programmers can use atomic sections to update the shared state. An atomic section is a small code sequence that nesC ensures will run automatically. TinyOS turns off interrupts during the atomic section and ensures that it has no loops.

6.7.1.3 Scheduling

TinyOS uses a non-preemptive schedule to place lengthy tasks that require lightweight dispatching into an interrupt context. TinyOS executes only one program consisting of selected system components and custom components needed for a single application. A complete system configuration consists of a tiny scheduler and a graph of components. It runs in a single address space

---

[1]First in, first out.

and contains two execution environments: interrupt handlers running at high priority and tasks that are scheduled in FIFO order. TinyOS adopts the two-level concurrent models based on the combination of tasks and events. Tasks provide a level of deferred processing, reserving interrupt contexts for short, time-critical sections of code. Tasks are equal and there is no concept of priority and no preemption between tasks. All tasks share one executing space, which saves the memory overhead in runtime. Tasks are managed by a circular queue and the task scheduling follows FIFO mode. Resources are distributed beforehand, and currently there can only be seven waiting tasks in the queue. The task-processing model is shown in Figure 6.13; eight tasks are in the queue. There are three tasks in the queue. If the task queue is null and there are no events occurring, the processor will enter into sleep mode automatically, and will be awakened by hardware interruption subsequently. This is conducive to saving energy [23]. Events are generated by hardware interruption (MCU external interruptions, timer interruptions, etc.) directly or indirectly. When receiving an event, TinyOS will execute the handler corresponding to the event immediately. The handler can preempt the current running task and this can be applicable to time-critical application. It is an asynchronous, time-response fast executive mode.

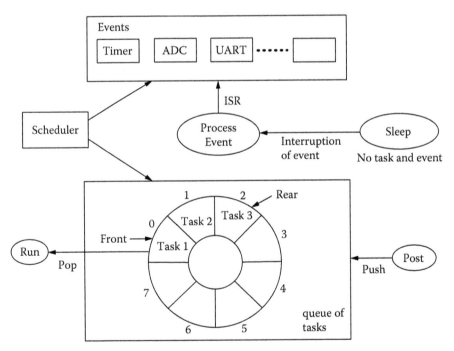

FIGURE 6.13
Scheduling strategy of TinyOS [23]

6.7.2 Contiki

Contiki [3], developed at the Swedish Institute of Computer Science, is an event-driven operating system that has very minimal memory requirements. Contiki uses a hybrid model to combine the benefits of both event-driven systems and threads. The system is based on an event-driven kernel where multithreading is implemented as an application library. Contiki is suitable for embedded systems, where memory management hardware such as an MMU is not available on the platform. This is common for microcontrollers as well as low-end processors. The kernel and applications in Contiki share the same memory space due to the memory and processing constraints of the underlying environment. To make an operating system useful for the WSN environment, it must not consume all the resources of the underlying platform. Therefore, one of the driving forces behind the Contiki operating system is to minimize the resources consumed. One downside of this design is that the kernels of Contiki are not protected from applications, and therefore have access to each others' global data. Moreover, an application also exists in the same memory space as another application. Therefore, applications can also corrupt each other in addition to corrupting the kernels. However, the fact that the applications and kernel share memory space makes it very efficient for them to share information through global variables. In spite of the fact that this causes the operating system and applications to be tightly integrated, it allows for more efficient usage of the available memory. Due to this approach, typical configurations of the Contiki operating system can fit within 2 K of RAM and 40 K of ROM.

6.7.2.1 Kernels

Contiki consists of the kernel libraries, the program loader and a set of processes. A process may be an application program or a service. Each application essentially is one process with a main entry point corresponding to an event handler. All interactions are based on either shared global data or event communication. Communication between processes always goes through the kernel. The kernel does not provide a hardware abstraction layer. However, it allows device drivers and applications to communicate directly with the hardware.

A Contiki system is partitioned into two parts: the *core* and the *loadable programs* as shown in Figure 6.14 [3]. This partitioning occurs at the compile time and is specific to the deployment requirements of the application in which Contiki is used. The core consists of the Contiki kernel, the program loader, the most commonly used parts of the C language runtime and support libraries, and a communication stack with device drivers for the communication hardware. The core is compiled into a single binary image that is stored in the devices prior to deployment. The core is generally not modified after

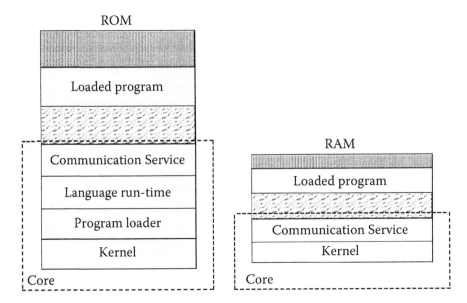

FIGURE 6.14
Partitioning into core and loadable programs [3]

deployment. Loadable programs are installed on top of the core and do not modify the core. Programs are loaded into the system by the program loader.

The Contiki *kernel* consists of a lightweight event scheduler that dispatches events to running processes and periodically calls processes' polling handlers. All program execution is triggered either by events dispatched by the kernel or through the polling mechanism. Normally, the kernel will only send events to a process when it is created (INIT event), stopped (EXIT and EXITED events) or polled (POLL event), whereas processes can send events whenever they need, depending upon the implemented functionality corresponding to the process. An event cannot be received by the kernel. A process event handler is the only place where events can be received. The kernel does not preempt an event handler once it has been scheduled. Therefore, event handlers must run to completion, which is common in event-based operating systems. This means that no preemption is employed by the operating system. Alternatively, the processes are expected to quickly handle received events, register to receive new events and finally, return from the event handler so that another process can be scheduled.

The kernel supports two kinds of events: *asynchronous* and *synchronous* events. Asynchronous events are deferred procedure calls when an event is not immediately delivered. In this case, events are queued in the kernel's event queue and delivered when the OS schedules that event to be delivered. Since the event queue is a FIFO activity, events will be scheduled based on the order in which they were added to the queue. Event handler routine is the main entry point for a process. To receive an event, the receiver process (or processes) invokes the event handler routine with an argument specifying the event to be sent. On the other hand, synchronous events cause the events to be scheduled immediately and the receiver process (or processes) should also invoke their event handler. As this is a blocking call, control returns to the initiator process only after the target process has finished the event. The Contiki kernel uses a single shared stack for all process execution. The use of asynchronous events reduces stack space requirements as the stack is rewound between each invocation of event handlers.

In addition to the events, the kernel provides a polling mechanism. Polling is a way to signal a process to do something. It is normally associated with interrupt handlers. Processes that operate near the hardware to check for status updates of hardware devices can use polling. Interrupt handlers are typically associated with device drivers to service I/O requests of the supporting hardware. In this situation, a process may need to use polling and the polling activity is dependent on the type of the device. Polling is useful for high priority events that are scheduled in-between each asynchronous event. When a poll is scheduled all processes that implement a poll handler are called in order of their priority. Using the POLL API, a polling flag is set in the process structure of an associated I/O process. The kernel scheduler checks this flag and triggers a POLL event be posted to the associated process. That process will then perform whatever data operations might be associated with a POLL operation.

6.7.2.2 Loadable Programs

Programs to be loaded into the system are generally stored in EEPROM before they are loaded into the code memory. Loadable programs are implemented using a runtime relocation function and a binary format that contains relocation information. When a program is loaded into the system, the loader first tries to allocate sufficient memory space based on information provided by the binary. The program loader can obtain the program binaries either by using the communication stack or directly using the attached storage. If memory allocation fails, program loading is aborted. After the program is loaded into memory, the loader calls the program's initialization function. The initialization function may start or replace one or more processes. The core has no information about the loadable programs, except for those that explicitly register with the core. On the other hand, loadable programs have full knowledge of the core and can call functions as well access variables that reside in the core. Loadable programs can call each other by going through the kernel.

A loadable module contains the object file of the program that is to be loaded into the system. In dynamic linking, the object files contain code, data, function name and variables of the system core that are referred by the module. However, the code in the object file cannot be executed before the physical addresses of the referenced variables and functions are assigned. This process is done at runtime by a dynamic linker. ELF (executable and linkable format) and compact ELF are two file formats for dynamically linked modules used in the Contiki dynamic linker [8]. The Contiki core contains a table of the symbolic names of variables and function names as well their addresses. The symbol table is used by the dynamic linker when linking loaded programs. The dynamic linker reads ELF/CELF files through the Contiki virtual file system interface. As a result, the dynamic linker does not need to know the physical location of the ELF/CELF file and can operate on files stored in RAM, flash-ROM and external EEPROM. The dynamic linker performs four steps to link, relocate and load an ELF/CELF file.

- `Information collection`: The dynamic linker parses the ELF/CELF file and extracts relevant information. This information mainly concerns where in the ELF/CELF file the code, data, symbol table, and relocation entries are stored. The relocation information in an ELF/CELF file consists of a list of relocation entries. A relocation entry contains a pointer to a symbol and corresponds to an instruction or address in the code or data in the module that needs to be updated with a new address.

- `Memory allocation`: Memory for the code and data is allocated from flash ROM and RAM, respectively. Memory allocation for the loaded program is done using a simple block allocation scheme.

- `Linking and relocating`: The code and data segments are linked and relocated to their respective memory locations. The dynamic linker processes one relocation entry at a time. Its symbol is looked up in the symbol table in the core. If it is found, the address of the symbol is used to patch the code or data. If it is not found in the symbol table of the core, then the symbol table of the ELF/CELF file is searched.

- `Loading`: After the linking and relocating, the code and the data are loaded into their final memory location. The code is written to flash ROM and the data to the RAM.

Contiki starts executing the program once the dynamic linker has successfully loaded the code and data segments. The dynamic linker is designed to be easily portable between different platforms and is divided into two modules:

- `Generic module`: This module parses and analyzes the ELF/CELF to be loaded.

- `Architecture-specific module`: This module allocates ROM and RAM for the program to be loaded, performs code and data relocation, and writes the linked and relocated binary to flash ROM.

However, dynamic linking of ELF requires support from the underlying operating system and cannot be done on a monolithic operating system.

Another approach is JVM by which Java programs can call native code methods by declaring native Java methods. The JVM dispatches calls to native methods to native code. In this way, any native function can be called in Contiki including services within a loaded Contiki program. Contiki virtual machine (CVM) is designed as a stack-based machine with separated code and data areas. CVM can be configured for the application running on top of the machine by allowing functions to be implemented as native code or as CVM code. A CVM program can call any native function provided by the underlying system through the native function interface. Invocation of native functions is done through a special instruction which takes one parameter to identify which native function to be called. Native functions in a CVM program are invoked like any other normal function.

According to [5], a combination of native and virtual machine code is an energy-efficient alternative to pure native code or pure virtual machine code approaches. Therefore, the dynamic linking mechanism can be used to load the native code used by the virtual machine code in the virtual machines.

6.7.2.3   Services

A service in Contiki system is nothing but a process like a shared library. Generally, it implements functionality that can be used by other processes if required services can be dynamically replaced at runtime and also be dynamically linked. Communication protocol stacks, device drivers, and higher level algorithms are examples of services. Services are managed by a service layer that contains fundamental system services for applications. This layer contains the service descriptions and also keeps track of running services. A service consists of a service interface that defines the methods to be exposed as a service and a process that implements that interface and handles actual servicing of incoming service requests. The service interface consists of a version number and a function table with pointers to the functions that implement the interface. A stub library is used by the application program to access the service. The stub library uses the service layer to find the required service process and then caches the process identification for future requests. The kernel in the Contiki system only provides the basic CPU multiplexing and event handling features. The major part of the system is implemented as system libraries that can be linked with applications in three different ways. Figure 6.15 shows an application function calling a service.
The three linking methods are:

- Static linking with core libraries

- Static linking with loadable program libraries

- Call services that implement a specific library

One of the important service in Contiki is communication. Communication services use the service mechanism to call each other and synchronous events

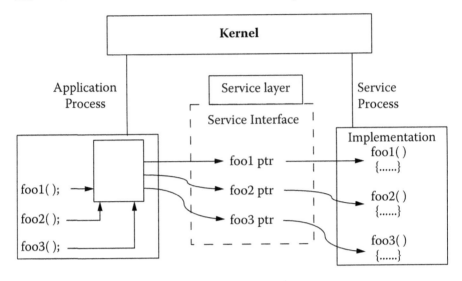

FIGURE 6.15
An application function calling a service in Contiki [3]

use the mechanism to communicate with application programs. Additionally, the communication stack can be split into different services as shown in Figure 6.16.

6.7.2.4 Protothreads

Contiki applications run as extremely lightweight protothreads with a very small RAM overhead [6] [7]. Protothreads are programming abstractions that provide a conditional blocking wait operation and can be used to reduce the number of explicit state machines in event-driven programs for memory-constrained devices. Each application essentially is one process with a main entry point corresponding to an event handler. All interactions are based on shared global data or event communication. In a multithreading system each thread requires its own stack. Generally, in memory-constrained systems, memory must be statically reserved for the threads. This reserved memory cannot be used for other purposes, even when a thread is not currently executing. In contrast to multithreading, protothreads are stackless. That means they do not have a history of function calls. All protothreads run on the same stack and context switching is done by stack rewinding. Protothreads can be seen as a combination of events and threads. Protothreads have inherited the blocking wait semantics from threads and the stackless property and the low memory overhead from events. The minimum stack memory requirement is therefore the same as the maximum stack usage of all programs, whereas,

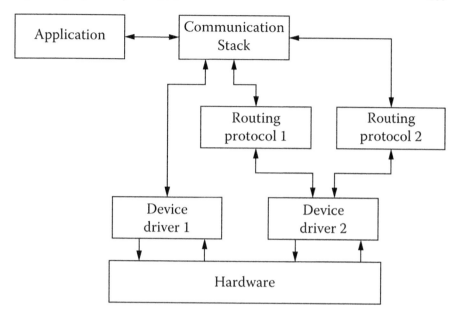

FIGURE 6.16
Loosely coupled communication stack in Contiki [3]

the minimum memory requirement for stacks in multi-threaded systems, is the sum of the maximum stack usage of all threads. Figure 6.17 shows the stack memory requirements for three event handlers written with protothreads and equivalent functions running in three threads. Event handlers and protothreads run on the same stack and each thread runs on a stack of its own.

Programs written for an event-driven model typically have to be implemented as explicit state machines. Programs with protothreads can be written in a sequential fashion without having to design explicit state machines. A protothread consists of two main elements, an event handler and a local continuation. The event handler is associated with the process. A local continuation is used to save and restore the context when a blocking protothread API is invoked. It is a snapshot of the current state of the process.

In Contiki, protothreads are implemented using C programming language and without any architecture-specific machine code. Protothread APIs are implemented as macros. When these macros are expanded, they expose the implementation of the protothreads. A protothread runs only within a single C function and cannot span over other functions. It can also call normal C functions, but cannot block inside a called function. Protothreads are based on a low level mechanism called local continuation which is similar to ordinary continuation, but does not capture the program stack [7]. Continuation is an abstract representation of the control state of a program. In other words, it

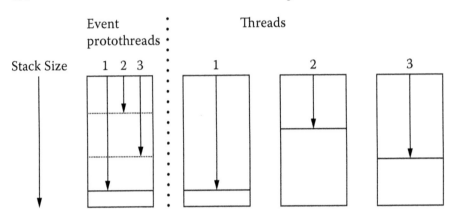

FIGURE 6.17
Stack requirements: protothreads versus threads [7]

is a data structure that represents the computational process at a given point in the execution. A protothread consists of a C function and a single local continuation. A local continuation supports two operations namely, **set** and **resumed**.

- The **set** operation of a local continuation captures the state of the function, all CPU registers including program counter but not the stack.

- The **resumed** operation of that same local continuation resets the state of the function to what it was initially, i.e., when the local continuation was set.

In spite of the fact that protothreads allow programs to take advantage of a number of benefits of the threaded programming model, they also impose some of the limitations from the event-driven model. There are a number of limitations associated with local continuations. Some of them are highlighted here.

- Preemption is not possible in event-driven models or protothreads as a single stack is used. According to the semantics of the threaded model, the thread's stack is saved when it is preempted and execution of another thread is continued.

- Automatic variables are not saved across a blocking wait in event-driven models and protothreads. Although automatic variables can still be used inside a protothread,the contents of the variables must be explicitly saved before executing a wait statement because both event-driven models and protothreads use a single shared stack for all active programs, and rewind the stack every time a program blocks.

Preemptive multi-threading in Contiki is implemented as a library on top of the event-based kernel of the system and it provides an API to applications that must run preemptively in a multithreaded mode. In other words, it provides optional preemptive multithreading that can be applied to individual processes and these routines are called explicitly by that process to run and schedule threads.

The library has two parts. A platform-independent part that interfaces to the event kernel and a platform-specific part for implementing the stack switching and preemption primitives. Preemption is implemented using a timer interrupt that saves the processor registers onto the stack and switches back to the kernel stack. This is not the case of normal Contiki processes that share a common stack. The stack management functions are provided by the library. The API of the multithreading library is shown in the table below. Functions `mt_start()` and `mt_exec()` are called to setup and run a thread. An event handler can call a function `mt_exec()` to schedule a thread. Functions `mt_yield()`, `mt_post()`, `mt_wait()`,and `mt_exit()` can be called from a running thread. The Contiki OS employs a lightweight scheduler that dispatches processes when there are events posted against them or when the polling flag is set. Once a process is scheduled and the event handler starts executing, the kernel does not preempt it until completion.

API	Description
mt_start(thread, functionptr, dataptr);	Start thread with specified function call.
mt_exec(thread);	Execute specified thread until it yields or is preempted.
mt_post(id, event, dataptr);	Post event from running thread.
mt_wait(event, dataptr);	Wait for event to be posted to running thread.
mt_yield();	Yield from running thread
mt_exit();	Exit running thread.

6.7.2.5   Power Management

A set of mechanisms is provided by the Contiki operating system for reducing the power consumption of the underlying device on which it runs. Contiki MAC is the default mechanism for attaining low-power operation of the radio. In Contiki MAC, nodes can be running in low-power mode and still be able to receive and relay radio messages. The mechanisms in Contiki MAC are inspired by existing duty cycling protocols, such as B MAC, X MAC, and BoX MAC. Despite energy consumption at nodes depending on many factors (e.g., sensing, processing and communication), the radio subsystem is typically the most energy-consuming component. Hence, duty cycling is a widely used approach for reducing energy consumption and prolonging the life of the network. The wireless transceiver has the highest power consumption among all other components. The power consumption for reception is the same when a sensor overhears the packets not destined for it. To save power, a sensor's transceiver must be required to be completely turned off. In fact, transmitters have to wait for receiver wakeups (duty cycle) before starting data packets transmissions. A duty cycling mechanism is used to turn the transceiver on. It is a radio duty cycling protocol that uses periodical wakeups to listen to packet transmissions from neighbors.

Nodes sleep most of the time and periodically wake up to check for radio activity. If a packet transmission is detected during a wakeup, the receiver is kept on to receive the packet and then sends a link layer acknowledgement after receiving a packet successfully. To send a packet, the sender repeatedly sends the same packet until a link layer acknowledgment is received. To transmit a packet, a sender repeatedly sends its packet during the wakeup interval to ensure that all neighbors have received it. The sender continues sending packets until it receives a link layer acknowledgement from the receiver as shown in Figure 6.18. Packets sent as broadcasts do not result in link layer acknowledgments. Instead, the sender repeatedly sends the packet during the full wakeup interval to ensure that all neighbors have received it as shown in Figure 6.19.

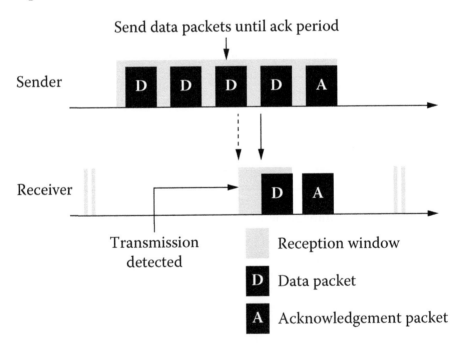

FIGURE 6.18 (SEE COLOR INSERT)
Duty cycle in Contiki MAC [2]

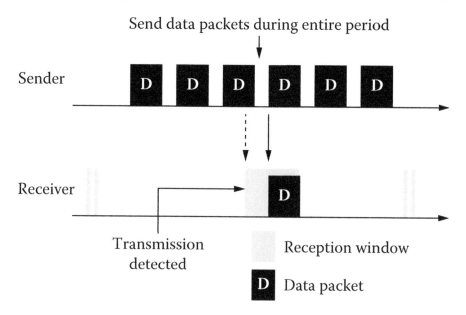

FIGURE 6.19 (SEE COLOR INSERT)
Broadcast transmission in Contiki MAC [2]

A wakeup mechanism in Contiki MAC uses a clear channel assessment (CCA) scheme that employs the received signal strength indicator (RSSI) of the radio transceiver to give an impression of radio activity on the channel. The CCA returns a positive value if the RSSI is below a certain threshold, which means the channel is clear. If the channel is in use, the RSSI will be above the threshold and the CCA will return a negative value. However, CCAs in Contiki MAC cannot detect package transmissions reliably. They can only detect RSSIs above a certain value. Figure 6.20 shows the Contiki MAC transmission and CCA timing.

Wake-up mechanism in ContikiMAC uses Clear Channel Assessment (CCA) scheme that employs Received Signal Strength Indicator (RSSI) of the radio transceiver to give an impression of radio activity on the channel. The CCA returns positive value, if the RSSI is below a given threshold, which means the channel is clear. If the channel is in use, the RSSI will be above the threshold and CCA will return negative value. However, CCAs in ContikiMAC can not detect packet transmissions reliably. They can only detect that the radio signal strength (RSSI) above a certain value. Figure 6.20 shows the ContikiMAC transmission and CCA timing.

The timing must satisfy following constraints:

- The interval between each packet transmission t_i must be smaller than the interval between each CCA t_c. This is to ensure that either the first

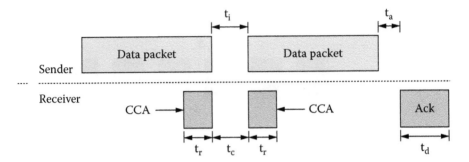

FIGURE 6.20
Contiki MAC transmission and CCA timing [2]

or the second CCA is able to see the packet transmission. If t_c is smaller than t_i, two CCAs would not be able to reliably detect a transmission.

- A packet transmission cannot be so short that it falls between the CCAs. Specifically, t_s, the transmission time of the shortest packet, must be larger than $t_r + t_c + t_r$.

There may be different reasons behind detecting a radio signal by a receiver's CCA. It may that a neighbor is sending a packet to that receiver or sending a packet to some other receiver that is overheard. If a node is transmitting a packet to the receiver, the receiver should stay awake to receive the full packet and after receiving it should transmit a link layer acknowledgement. In contrast, if a node overhears a packet it should quickly go into sleep mode again. Contiki MAC implements fast sleep optimization and transmission phase-lock techniques. The fast sleep optimization technique allows receivers to go back to sleep if a transmission is not intended for it. The receiver goes back to sleep:

- If a silence period is not detected before the transmission time

- If a silence period is longer than the interval between each packet transmission

- If no packet is received after the silence period, even though radio activity is detected

A sender can optimize its transmission by predicting receiver's wakeup phase. This can be learned from the time at which a sender receives a link layer acknowledgement from the receiver after a successful transmission. After a sender has learned the wakeup phase of a receiver, it can send its future transmissions at this receiver just before the receiver is expected to be awake. This mechanism is called transmission phase lock. The phase-lock mechanism maintains a list of neighbors and their wakeup phases. The phase-lock mechanism is implemented in Contiki as a separate module from Contiki MAC, to allow it to be used by other duty cycle mechanisms, for example, Contiki X MAC.

The Contiki MAC implementation uses the Contiki real-time timers to schedule periodic wakeups. The real-time timers can preempt any Contiki process at the time at which they are scheduled. The wakeup mechanism of Contiki MAC runs as a protothread. These protothreads perform the periodic wakeups and also implement the fast sleep optimization technique. Contiki initializes a few processes at a regular start-up and calls the system function `process_run()`. This function calls all registered poll handlers and then processes one event from the system event queue. After the single event is processed, the function returns the number of unattended events still in the queue. When there is no event to process, the system may choose to sleep to save energy. The system can be active again if there is any external interrupt and it loops around the `process_run()` function until all new events are handled.

6.7.2.6 Networking

Contiki and its uIPv6 stack constitute the first low-power wireless operating system to provide an IPv6 ready stack [16] [4]. Contiki provides an implementation of uIP, a TCP/IP protocol stack for small 8-bit microcontrollers. The uIP implementation has the minimum set of features needed for a full TCP/IP stack and supports TCP, UDP, ICMP, and IP protocols. uIP uses a single global buffer for holding packets, large enough to contain only one maximum-sized packet. When a new data packet arrives from the network, the network device driver puts it in the global buffer and calls uIP to handle the new data. uIP notifies the intended application after analysing the incoming packet. The application must act quickly to avoid the data being overwritten by another incoming packet because they all use the same buffer. In some cases, the application may also choose to immediately send a response using the same global buffer. An application can call the network device driver to send the packet. The IPv6 stack for low-power wireless is shown in Figure 6.21. The stack consists of the standard IPv6 protocols at the network layer, transport layer and a set of new protocols from network layer and down.

Contiki also provides another lightweight layered communication stack, called Rime,[2] for sensor networks to simplify the implementation of communication protocols [1]. Layers in Rime are very thin which makes it different from traditional network architecture. The main purpose of Rime is to simplify implementation of sensor network protocols and facilitate code reuse. The thin layers enable code reuse within the stack. Rime uses a single buffer for both incoming and outgoing packets similar to uIP and that helps reduce its memory footprint. Rime is a part of Contiki's system core, which is present

---

[2]Available in Contiki 2.6 and higher version.

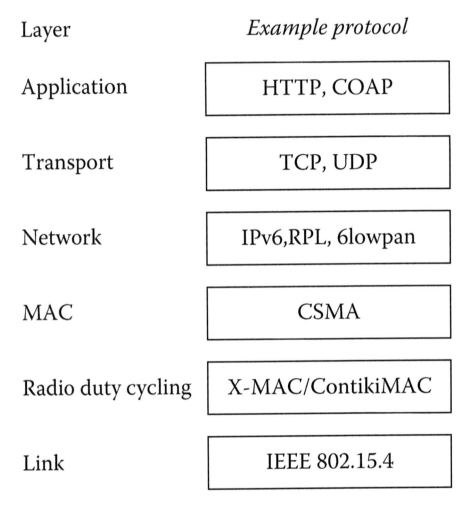

Layer	*Example protocol*
Application	HTTP, COAP
Transport	TCP, UDP
Network	IPv6,RPL, 6lowpan
MAC	CSMA
Radio duty cycling	X-MAC/ContikiMAC
Link	IEEE 802.15.4

FIGURE 6.21
IPv6 stack for low-power wireless [4]

in memory. Thus, Rime takes the burden of memory footprint and loadable programs are made smaller. Latest versions of Contiki contain Contiki RPL, which is an implementation of routing protocol for low-power and lossy networks (RPL) routing [17]. RPL is the IETF-proposed standard protocol for IPv6 routing over low power lossy networks (LLNs). In RPL, sensors choose parents based on an objective function, for example, optimizing power consumption or optimizing latency can be an objective function. Moreover, RPL allows individual networks to choose different objective functions. The default objective function in RPL is called Objective Function 0 (OF0), which optimizes hop count. Contiki RPL implements the RPL protocol according to RPL specification Version 18. It implements two objective functions namely, OF0 and the minimum rank objective function with hysteresis (MRHOF). The implementation separates protocol logic, message construction and parsing and objective functions into different modules. RPL is designed for running over radio layers such as IEEE 802.15.4 which has lower output power than radios such as WiFi and Bluetooth.

Contiki RPL is implemented in C using the APIs of the Contiki operating system. More than one RPL instance can run in a network. Each RPL instance defines a topology and multiple directed acyclic graphs (DAGs) may exist, each having a different directed acyclic graph (DAG) root. A sensor node must belong to one DAG in each instance. However, it can also join multiple RPL instances. During the formation of the topology, each sink generates a packet called a DAG information object (DIO), and sends it to all children. Any node that decides to join the DAG may pass the DIO further to its own children. This process continues until all the nodes decide their DAG for that RPL instance. The DIO also contains a rank to indicate the size of the DAG. When a node joins the DAG it increases the rank. The rank can be utilized to distinguish between parents and siblings and help prevent routing loops. For the maintenance of the topology, DAGs may need adjustment which is initiated by the root. A trickle timer is used for this purpose to decide when nodes should forward such information to their children. To provide support for point-to-point communication, a destination advertisement object (DAO) is included in RPL. This allows a node to advertise its address prefixes among its neighbors using multicast and also communicate with the DAG root using unicast after joining a DAG.

The routing protocol with RPL uses Contiki's modular IPv6 routing interface that has three functions: activate, lookup and deactivate.

- `Activate`: To initialize the DAG construction, a DAG information solicitation (DIS) is sent by the activate function. Neighbors that belong to a DAG will reset their trickle timers and soon thereafter the node will receive at least one DIO.

- `Lookup`: The lookup function is used for finding a route to a destination. If uIPv6 detects that the destination for a packet is not an immediate neighbor, it sends a route lookup request to the RPL.

- `Deactivate`: The deactivate function is used to deallocate the internal structures and sends no-DAO message to the node's neighbors. The module also stops responding to route lookup requests after deactivation. However, it may be reactivated later.

6.7.3 MagnetOS

MagnetOS [12] [20] [14], developed at Cornell University, is a distributed operating system for ad hoc and sensor networks. This system implements a network-wide, energy-efficient virtual machine on top of a collection of ad hoc nodes. MagnetOS is based on JVM, provides access to standard Java libraries and enables development tools to construct distributed applications for WSN. The goals of MagnetOS are as follows [14]:

- Efficiency: Efficient execution model of network applications can increase system lifetime by reducing excessive communication or power consumptions.

- Adaptation: Dynamic adaptation to the significant changes in the network makes the system robust.

- Generality: A wide range of general purpose applications should be supported by the system. New applications should be developed with little effort.

- Extensibility: The system must be designed so that implementation takes future growth into consideration.

- Compatibility: The system should be compatible with the existing development tools.

The idea of MagnetOS is to satisfy these goals and make the network operate as an extended JVM so that the entire network appears as a single unified system to applications. MagnetOS provides this single-system image programming model based on events. An event is an indication that a piece of code should be executed in response to its occurrence. Both synchronous and asynchronous events can be handled by MagnetOS. In case of synchronous events, the event invoker blocks until termination of the handler. In asynchronous events, the control immediately returns to the component that invoked the event. Applications are structured as a set of interconnected, mobile event handlers specified statically as objects. Monolithic Java applications consist of a set of event handlers, and execution consists of a set of event invocations that may be performed concurrently. The communication and migration of these application segments among the nodes in the network are then coordinated by the MagnetOS runtime.

6.7.3.1   Partitioning and Migration

MagnetOS has static and dynamic components to manage partitioning. Static partitioning extracts from each original application class an event handler, a dispatch handle, an event descriptor and a set of event globals associated with the event handler [14]. The event globals are static fields shared across all instances of an event handler. Each event handler retains pointers to the corresponding instance of event globals, and can therefore share state with other handlers. Each handler is free to move across nodes in the network. Dispatch handles are used to invoke procedure calls on remote event handlers residing on other nodes. This is implemented by a remote procedure call (RPC) and enables code migration.

The dynamic services provided by the MagnetOS runtime make possible the distributed execution of components of an application across the network. Dynamic services include component creation, inter-component communication, migration, garbage collection, naming, and event binding.

An application contacts the local runtime to create a new instance of an event handler and passes the requisite type descriptor and parameters for creation. The runtime then decides a suitable location within the network for the placement of the newly created handler. Each runtime keeps a list of local event handlers. The current location of the handler is maintained by the dispatch and processes call events on behalf of application invocations by marshalling and unmarshalling of event arguments and results. MagnetOS also takes into account the migration overhead while making a decision to relocate a handler. Migration is invoked to preserve the computation without disruption. This is required in some situations, when the battery level of a node is below a critical threshold. Here it is required to offload all event handlers at that node. Another suitable node is selected for the rest of the execution. MagnetOS runtime transports the state and resumes the computation at the destination.

6.7.3.2 API Abstraction

MagnetOS provides some important abstractions for programmers to be able to implement sensor network applications effectively. These abstractions enable applications to name nodes and application components, to collect statistics and information on network and node behavior, and to set and direct migration policies. Some of the important parts of API are discussed here.

- A `Node` encapsulates the notion of a physical sensor that is running a MagnetOS instance. It enables applications to query node properties such as link state, network topology information and energy status.

- The `Link` abstraction refers to a physical link between two sensors, i.e., `Node` and the `NeighborSet` abstraction is the set of one-hop neighbors of a given sensor. These two abstractions enable applications to discover the network topology and link characteristics based on low-level information updated by the operating system.

- The `Energy` API enables applications to query the current energy level, the drain rate, and recharge frequency of the underlying heterogeneous platforms through a common interface.

- `Timer` abstraction enables applications to schedule events to be invoked in future.

- `Lock` is a remote synchronization object analogous to the monitor in Java.

- A `Thread` in MagnetOS is an execution context. When distributed over multiple nodes it can migrate from one node to another. Each thread in MagnetOS has a unique identifier, which is essential for the remote synchronization in the system.

The main idea of energy efficient execution in MagnetOS is to move communication endpoints and shorten the path packet traverse in the network. Authors [14] claim that it increases energy utilization and extends network lifetime.

6.7.4   Mantis OS

MultimodAl system for NeTworks of In situ wireless Sensors (MANTIS) OS is an open source multithreaded cross platform embedded operating system for WSN [13] [21]. Mantis OS (MOS) was developed as a part of the Mantis project by the research group at the University of Colorado, Boulder. MOS addresses the resource constraints of wireless sensor networks, namely limited memory and power. MOS is implemented in a lightweight RAM footprint which includes kernel, scheduler and network stack and fits in less than 500 bytes of memory. A power-efficient scheduler is implemented as a part of the MOS to achieve energy efficiency. Multithreading and energy efficiency are not mutually exclusive. MOS exploits that feature by using a multithreaded system to sleep efficiently when application threads indicate that there is no useful work to done. The architecture of the MANTIS operating system is shown in Figure 6.22.

The key design issues of MOS are ease of use and flexibility. The design of MOS is based on the classical structure of layered multithreaded operating systems and includes multithreading, preemptive scheduling with time slicing, I/O synchronization via mutual exclusion, a standard network stack, and device drivers. Moreover, preemptive multi-threading in the MOS enables sensor nodes to natively interleave complex tasks with time-sensitive tasks. Knowledge about these classical structures is sufficient for mastering this system which is very helpful for developers. MOS is written in standard C. Thus, MOS has the ability to shorten development cycles by enabling prototyping of applications and allows quick testing and debugging of new modules in the MOS kernel.

6.8   Summary

Operating systems for wireless sensor networks are discussed in this chapter. Operating systems for WSNs are different from traditional OSs. This chapter presents OS concepts in such a way that readers will learn about the interplay of WSN hardware and software via an OS acting as an interface between

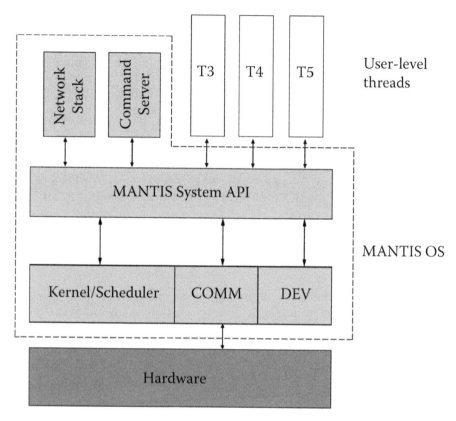

FIGURE 6.22
MANTIS OS architecture [21]

them. The chapter begins by discussing major design issues in operating systems for WSNs and covers architecture, execution models, scheduling, power management and communication concepts. Some popular operating systems for sensors like TinyOS, contiki, MagnetOS and Mantis OS are discussed in detail.

Bibliography

[1] Dunkels A. Rime: a lightweight layered communication stack for sensor networks. In *Proceedings of the European Conference on Wireless Sensor Networks (EWSN)*, 2007.

[2] Dunkels A. The Contiki MAC radio duty cycling protocol. In Technical Report T2011:13, Swedish Institute of Computer Science, 2011.

[3] Dunkels A., Grönvall B., and Voigt T. Contiki: a lightweight and flexible operating system for tiny networked sensors. In *Proceedings of First Workshop on Embedded Networked Sensors, (Emnets-I)*, IEEE, 2004.

[4] Dunkels A., Eriksson J., and Tsiftes N. Low-power interoperability for the IPv6-based internet of things. In *Proceedings of Tenth Scandinavian Workshop on Wireless Ad hoc Networks (ADHOC 11)*, 2011.

[5] Dunkels A., Finne N., Eriksson J., and Voigt T. Reprogramming wireless sensor networks with run-time dynamic linking in contiki. ACM Press, 2006.

[6] Dunkels A., Schmidt O., and Voigt T. Using protothreads for sensor node programming. In *In Proceedings of Workshop on RealWorld Wireless Sensor Networks*, 2005.

[7] Dunkels A., Schmidt O., Voigt T., and Ali M. Protothreads: simplifying event-driven programming of memory-constrained embedded systems. In *Proceedings of Fourth Conference on Embedded Networked Sensor Systems, ACM*, 2006.

[8] Dunkels A., Schmidt O., Voigt T., and Ali M. Simplifying memory-constrained event-driven programming with contiki's protothreads. In *Proceedings of Real-Time*, 2007.

[9] Texas Instruments CC2420 datasheet. Website: `http://focus.ti.com/lit/ds/symlink/cc2420.pdf`.

[10] TelosB Datasheet Crossbow. Website: `http://www.xbow.com/Products/Product\_pdf\_files/Wireless\_pdf/TelosB\_Datasheet.pdf`.

[11] Gay D., Levis P., von Behren R., Welsh M., Brewer E., and Culler D. The nesC language: a holistic approach to networked-embedded systems. In *Proceedings of Programming Language Design and Implementation (PLDI)*, ACM, 2003.

[12] Sirer E. G., Barr R., Kim T. W. D., and Fung I. Y. Y. Automatic code placement alternatives for ad hoc and sensor networks. In Computer Science Technical Report 2001-1853, Cornell University, 2001.

[13] Abrach H., Bhatti S., Carlson J., Dai H., Rose J., Sheth A., Shucker B., Deng J., and Han R. MANTIS: system support for multimodal networks of in-situ sensors. In *Proceedings of 2nd International Workshop on Wireless Sensor Networks and Applications (WSNA)*, 2003.

[14] Liu H., Roeder T., Walsh K., Barr R., and E. G. Sirer. Design and implementation of a single system image operating system for ad hoc networks. In *Proceedings of International Conference on Mobile Systems, Applications, and Services (Mobisys)*, 2005.

[15] Hill J., Szewczyk R., Woo A., Hollar S., Culler D. E., and Pister K. S. J. System architecture directions for networked sensors. In *Proceedings of the Ninth International Conference on Architectural Support for Programming Languages and Operating Systems*, 2000.

[16] Durvy M., Abeill J., Wetterwald P., C. O'Flynn, Leverett B., Gnoske E., Vidales M., Mulligan G., Tsiftes N., Finne N., and Dunkels A. Making sensor networks IPv6 ready. In *Proceedings of Sixth Conference on Networked Embedded Sensor Systems*, ACM, 2008.

[17] Tsiftes N., Eriksson J., and Dunkels A. Low-power wireless IPv6 routing with Contiki RPL. In *Proceedings of 9th International Conference on Information Processing in Sensor Networks (IPSN 2010)*, 2010.

[18] Levis P., Madden S., Gay D., Polastre J., Szewczyk R., Woo A., Brewer E., and Culler D. The emergence of networking abstractions and techniques in Tiny OS. In *Proceedings of First USENIX/ACM Symposium on Networked Systems Design and Implementation (NSDI)*, 2004.

[19] Levis P., Madden S., Polastre J., Szewczyk R., Whitehouse K., Woo A., Gay D., Hill J., Welsh M., Brewer E., and Culler D. Tiny OS: an operating system for sensor networks. In *Ambient intelligence*, Springer, 2005.

[20] Barr R., Bicket J. C., Dantas D. S., Du B., Kim T. W. D., Zhou B., and Sirer E., G. On the need for system-level support for ad hoc and sensor networks. *Operating Systems Review*, 36(2):1–5, 2002.

[21] Bhatti S., Carlson J., Dai H., Deng J., Rose J., Sheth A., Shucker B., Gruenwald C., Torgerson A., and Han R. Mantis OS: an embedded multithreaded operating system for wireless microsensor platforms. *Mobile Networks and Applications*, 10(4):563–579, 2005.

[22] Raghunathan V., Schurgers C., Park S., and Srivastava M. B. Energy-aware wireless microsensor networks. *IEEE Signal Processing Magazine*, 19(2):40–50, 2002.

[23] Zhi-bin Z. and Fuxiang G. Study on preemptive real-time scheduling strategy for wireless sensor networks. *Journal of Information Processing Systems*, 5(3):135–144, 2009.

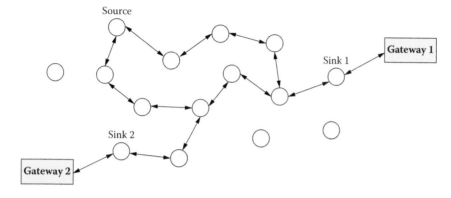

FIGURE 1.2
Multi-hop sensor network

FIGURE 1.3
Sensor motes: (a) Rene, (b) Mica2, (c) Mica2DOT

FIGURE 1.4
Sensor motes: (a) TelosB, (b) IRIS

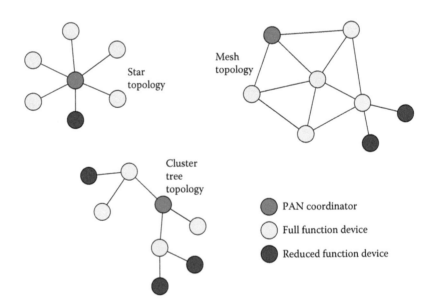

FIGURE 2.4
Three topologies of LR-WPAN

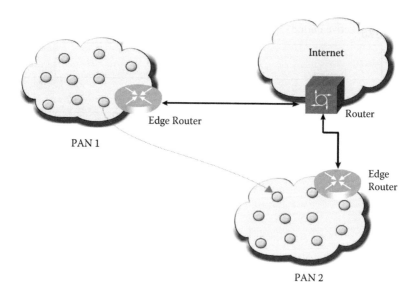

FIGURE 2.10
Personal area networks connected to Internet

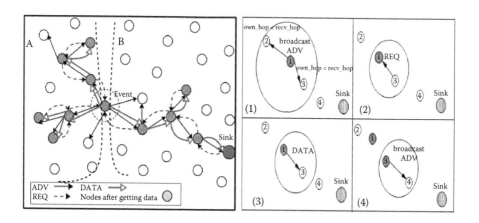

FIGURE 3.17
M-SPIN protocol in WSN: (a) partitioning into two regions, (b) negotiation using M-SPIN

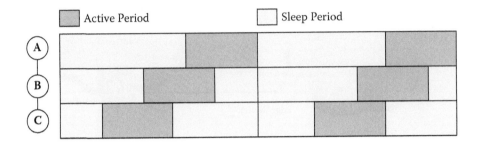

FIGURE 4.5
Staggered wakeup

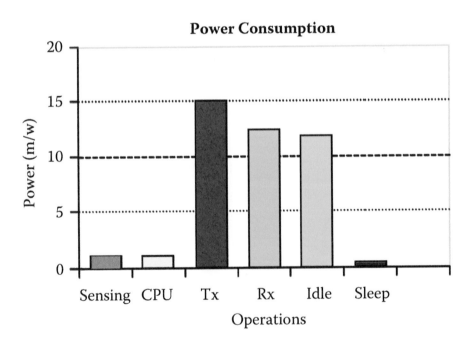

FIGURE 6.7
Energy consumption at different states

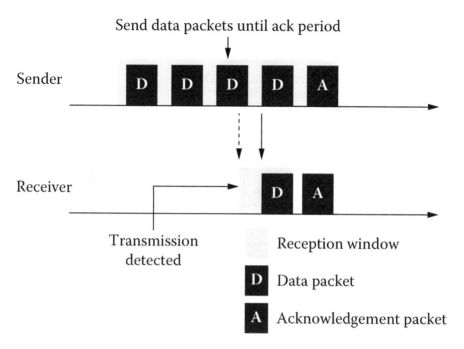

FIGURE 6.18
Duty cycle in Contiki MAC [2]

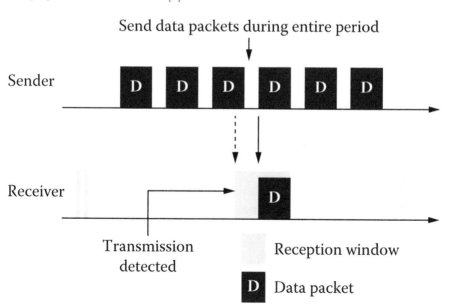

FIGURE 6.19
Broadcast transmission in Contiki MAC [2]

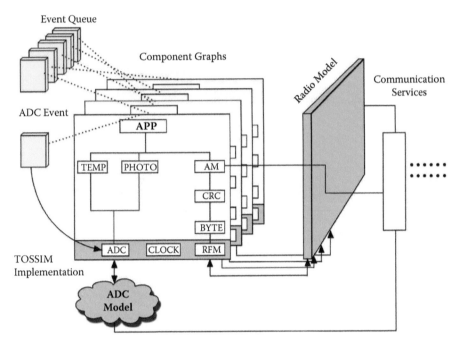

FIGURE 7.1
TOSSIM architecture [11]

FIGURE 7.2
Micaz mote connected to programming board

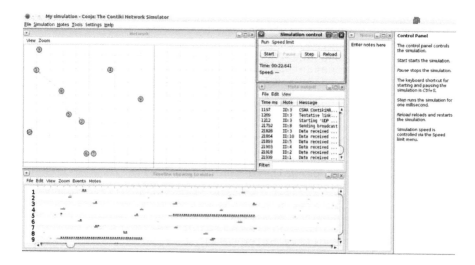

FIGURE 7.4
Screenshot of Contiki broadcast example

```
mobility.source = [
    'model/box.cc',
    'model/constant-acceleration-mobility-model.cc',
    'model/constant-position-mobility-model.cc',
    'model/constant-velocity-helper.cc',
    'model/constant-velocity-mobility-model.cc',
    'model/gauss-markov-mobility-model.cc',
    'model/hierarchical-mobility-model.cc',
    'model/mobility-model.cc',
    'model/position-allocator.cc',
    'model/random-direction-2d-mobility-model.cc',
    'model/random-walk-2d-mobility-model.cc',
    'model/random-waypoint-mobility-model.cc',
    'model/rectangle.cc',
    'model/steady-state-random-waypoint-mobility-model.cc',
    'model/waypoint.cc',
    'model/waypoint-mobility-model.cc',
    'model/sample-mobility-model.cc',
    'helper/mobility-helper.cc',
    'helper/ns2-mobility-helper.cc',
    ]
```

FIGURE 7.12
Entry of source file in wscript

```
headers.source = [
  'model/box.h',
  'model/constant-acceleration-mobility-model.h',
  'model/constant-position-mobility-model.h',
  'model/constant-velocity-helper.h',
  'model/constant-velocity-mobility-model.h',
  'model/gauss-markov-mobility-model.h',
  'model/hierarchical-mobility-model.h',
  'model/mobility-model.h',
  'model/position-allocator.h',
  'model/rectangle.h',
  'model/random-direction-2d-mobility-model.h',
  'model/random-walk-2d-mobility-model.h',
  'model/random-waypoint-mobility-model.h',
  'model/steady-state-random-waypoint-mobility-model.h',
  'model/waypoint.h',
  'model/waypoint-mobility-model.h',
  'model/sample-mobility-model.h',
  'helper/mobility-helper.h',
  'helper/ns2-mobility-helper.h',
]
```

FIGURE 7.13
Entry of header file in wscript

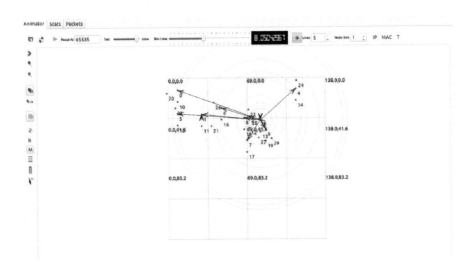

FIGURE 7.14
Packet animation in a wireless link

7

Programming WSNs

7.1 Introduction

This chapter discusses different programming techniques, especially designed to allow sensor network applications to be loaded onto the sensors deployed in the target area and forming a network using the wireless interface. In particular, WSN programming is a challenging task due to the limited resources of sensor nodes. Moreover, low data rate radio and dense networks result in slow transfer speeds of data and programs into a network. Thus effective programming techniques and platforms are required to build applications for WSNs.

Simulations of WSNs may facilitate the process of development of applications. They help examine the protocols, schemes, even new ideas designed for real environments. Moreover, WSN simulators allow users to isolate different factors by tuning configurable parameters. On the other hand, simulating a WSN is a complex process. As the processes in the simulators are often simplified, some technique which runs flawlessly in a simulator may fail in reality. In general, simulators designed for WSNs should provide support for wireless channel models (which is essential for the reliability of a communication network), energy consumption models (which influence the network lifetime), physical, MAC and network layer protocols. The range of available simulators for wireless networks is wide. Developing programs for WSNs in some popular platforms and simulators (TinyOS, COOJA, Castalia and NS-3) are discussed in this chapter.

7.2 TinyOS

The TinyOS SIMulator (TOSSIM) is discussed first. Then subsequent sections present the procedures for TinyOS installation and the implementation of protocols like the collection tree protocol (CTP) and modified SPIN (M-SPIN) in this platform.

7.2.1   TOSSIM

TOSSIM is a discrete event simulator specifically designed for WSNs running on TinyOS [10]. It was developed at the University of California at Berkeley as part of its TinyOS project. The TOSSIM framework is used mainly for testing and analysis. A WSN application for a mote can be compiled using a TOSSIM framework and runs on a PC. Programmers can easily examine their TOSSIM code using debuggers and other development tools. TOSSIM generates discrete-event simulations directly from TinyOS component graphs. It takes advantage of the TinyOS structure and whole system compilation to execute the same code that runs on sensor network hardware. TOSSIM translates hardware interrupts into discrete simulator events. The simulator event queue delivers the interrupts that drive the execution of a TinyOS application. Figure 7.1 shows the architecture of TOSSIM.

TOSSIM uses an abstraction for wireless networks. The network is represented as a directed graph, where each vertex is a node, and each edge has a bit error probability. To simulate asymmetric links, each edge (u, v) in the graph represents the error rate when mote u sends to v, and is distinct from the edge (v, u). TOSSIM provides mechanisms for TinyOS developers

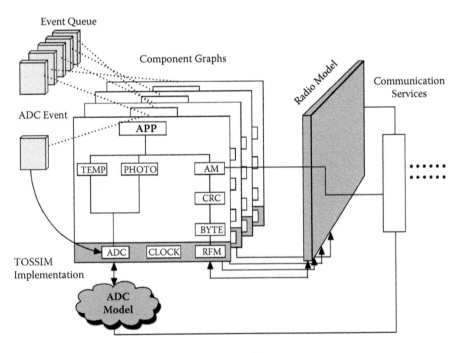

FIGURE 7.1 (SEE COLOR INSERT)
TOSSIM architecture [10]

to choose the accuracy and complexity of the radio model as necessary for their simulations. Each node has a private piece of state representing what it hears on the radio channel. Link probabilities can be specified by the user and changed at runtime. Transmission events propagate to the simulated input channel of each connected node. Each mote has its own local view of the network channel. This abstraction allows testing under perfect transmission conditions, i.e., when bit error rate is zero. This can also capture the hidden terminal problem and many of the different problems that can occur in packet transmission (start symbol detection failure, data corruption, etc.).

The TOSSIM simulator engine provides a set of communication services for interacting with external applications. These services allow programs to connect to TOSSIM over a TCP socket to monitor or actuate a running simulation. TOSSIM signals events to applications, providing data on a running simulation. Examples of events sent from TOSSIM are debug messages added by developers in TinyOS source code, radio and UART packets sent, and sensor readings. Applications call commands on TOSSIM to actuate a simulation and modify its internal state. Commands include operations to change radio link probabilities and sensor reading values, turn motes on and off, and inject radio and UART packets. Programs can also receive higher level information, such as packet transmissions and receptions or application-level events.

TinyViz is the TOSSIM visualization tool which can demonstrate the capabilities of TOSSIM communication services. TinyViz is a Java-based graphical user interface for visualizing simulations. It also provides visual feedback on the simulation state and mechanisms for controlling the running simulation, e.g., modifying ADC readings and radio loss probabilities. TinyViz provides a plugin interface to implement application-specific visualization and control code within its engine.

7.2.2 TinyOS Installation

The installation steps are as follows:

1. Remove any old tinyos repository from /etc/apt/sources.list and add deb http://tinyos.stanford.edu/tinyos/dists/ubuntu lucid main

2. Update the repository cache:

   $ sudo apt-get update

3. Run the following command and install the latest release of TinyOS:

   $sudo apt-get install tinyos

4. Add the following line in /root/.bashrc file to set up the environment for TinyOS at login: #Sourcing the tinyos environment variable setup script, then:

   source /opt/tinyos-2.1.2/tinyos.sh

5. Create a file named tinyos.sh in /opt/tinyos-2.1.2/ and add the following lines to the file:

    export TOSROOT=/opt/tinyos-2.1.2/

    export TOSDIR=/opt/tinyos-2.1.2/tos/

    export MAKERULES=/opt/tinyos-2.1.2/support/make/Makerules

    export CLASSPATH=$TOSROOT/support/sdk/java/tinyos.jar:.

    export PYTHONPATH=.:$TOSROOT/support/sdk/python:$PYTHONPATH

Blink is a very simple TinyOS application which flashes the LED lights. To check the installation, it can be compiled for either the mote hardware or for TOSSIM. To make the serial port writeable, change its mode:

  chmod 666 serialport

Here serialport is the serial port device name. For example, we can use

  /dev/ttySn-1

as the device name. Now connect the Micaz mote to the programming board as shown below and run following commands:

  $ cd /opt/tinyos-2.x/apps/Blink
  $ make micaz install mib510, serialport

If our installation is successful, we should observe counters on the three mote LEDs. In reality, it simply causes LED0 to turn on and off at 4 Hz, LED1 to turn on and off at 2 Hz, and LED2 to turn on and off at 1 Hz. The effect is as if the three LEDs were displaying a binary count of zero to seven every two seconds. Details of Blink application are available [13].

7.2.3 CTP in TinyOS

Collection is an essential operation to *disseminating* and *collecting* the data generated by a network into a base station. The main objective is to build one or more collection trees, each of which is rooted at a base station. When a sensor senses data which needs to be collected, it sends the data up the tree, and then forwards collection data that other nodes send to it. Implementation details of EasyCollection application in TinyOS are discussed in this section. In EasyCollection, nodes periodically send information to a base station which collects all the data. That is, when the send command is invoked, the data packet will be sent through the collection tree. Similarly, the receive event will only be called in the root of the tree, that is, in the base station. When the base

FIGURE 7.2 (SEE COLOR INSERT)
Micaz mote connected to programming board

station receives a *collected* packet, it just toggles a green LED. That means you have to install BaseStation in one TelosB mote which will be connected to the terminal and EasyCollection will be installed in other nodes. For example, take at least three motes. Now install this program into several nodes (make sure you have one base station, e.g., node with ID 1) and see how all the packets generated in the nodes are collected in the base station. The remaining nodes in the network periodically generate data and send it to the base station and toggle the blue LED.

1. Go to EasyCollection directory:

   $ cd /opt/tinyos-2.x/apps/EasyCollection

2. Check to see which serial port:

   $ motelist

3. Give write permission to the port:

   $chmod 666 usb-device-name

   or

   $chmod 666 /dev/ttyUSB0

   # connect the mote with USB serial port

4. To compile the code:

$make telosb

Installing:

$make telosb reinstall.2 /dev/ttyUSB0

Node ID is 1 for the root(base station here); for the others we can assign 2, 3 and so on.

EasyCollection is composed of two components: a module called EasyCollectionC.nc, and a configuration called EasyCollectionAppC.nc. All applications require a top-level configuration file, which is typically named after the application. In this case EasyCollectionAppC.nc is the configuration for the EasyCollection application and the source file that the nesC compiler uses to generate an executable file. EasyCollectionC.nc actually provides the implementation of the EasyCollection application. As you might guess, EasyCollectionAppC.nc is used to wire the EasyCollectionC.nc module to other components that the EasyCollection application requires. The reason for the distinction between modules and configurations is to allow a system designer to build applications from existing implementations. For example, a designer could provide a configuration that simply wires together one or more modules, which may not be his or her own design. Likewise, another developer can provide a new set of library modules that can be used in a range of applications. Sometimes (as is the case with EasyCollectionAppC and EasyCollectionC) a configuration and a module go together.

The nesC compiler compiles a nesC application when given the file containing the top-level configuration. Typical TinyOS applications come with a standard makefile that allows platform selection and invokes ncc with appropriate options on the application's top-level configuration. The following code shows the configuration for the EasyCollectionAppC.nc application:

```
configuration EasyCollectionAppC {}
implementation {
    components EasyCollectionC , MainC, LedsC , ActiveMessageC ;
    components CollectionC as Collector ;
    components new CollectionSenderC (0xee );
    components new TimerMilliC ();
    EasyCollectionC.Boot  -> MainC ;
    EasyCollectionC.RadioControl  -> ActiveMessageC ;
    EasyCollectionC.RoutingControl  -> Collector ;
    EasyCollectionC.Leds  -> LedsC ;
    EasyCollectionC.Timer  -> TimerMilliC ;
    EasyCollectionC.Send  -> CollectionSenderC ;
    EasyCollectionC.RootControl  -> Collector ;
    EasyCollectionC.Receive  -> Collector.Receive [0xee ];
}
```

The first thing to notice is the key word configuration, which indicates that this is a configuration file. Within the empty braces (configuration EasyCollectionAppC{}) it is possible to specify use and provide clauses, as with a module. This is important to keep in mind: a configuration can use and provide interfaces. In other words, not all configurations are top-level applications. A configuration is implemented within the pair of curly brackets following the key word implementation. The components lines specify the set of components that this configuration references. In this case, the components are Main, EasyCollectionC, LedsC, and three instances of a timer component called TimerMilliC which will be referenced as Timer. The remainder of the EasyCollectionAppC configuration consists of connecting interfaces used by components to interfaces provided by others. To fully understand the semantics of these wirings, it is helpful to look at the EasyCollectionC module's definition and implementation which will be discussed later. These wirings enable the LEDs and Timers to be initialized.

```
EasyCollectionC.Leds -> LedsC;
// Same as EasyCollectionC.Leds -> LedsC.Leds
```

Because EasyCollectionC only uses one instance of the Leds interface, this line would also work as:

```
EasyCollectionC -> LedsC.Leds;
// Same as EasyCollectionC.Leds -> LedsC.Leds
```

Similarly, the timer module as shown below is linked back in the module file as user interface TimerMilliC:

```
EasyCollectionC.Timer -> TimerMilliC;
```

The direction of a wiring arrow is always from a user to a provider. If the provider is on the left side, we can also use a left arrow:

```
TimerMilliC<- EasyCollectionC.Timer;
// Same as EasyCollectionC.Timer -> TimerMilliC;
```

For ease of reading, however, most wirings are left to right. Similarly, in this way, other modules and interfaces are connected together via wiring. To connect to boot sequence, the interface provided by MainC is used by the EasyCollection.Boot. Hence they are wired together. Another wiring is EasyCollectionC.RadioControl → ActiveMessageC, where ActiveMessageC gives basic communication interfaces such as:

```
configuration ActiveMessageC
Provides
interface AMPacket
interface AMSend[am_id_t]
interface Packet
interface PacketAcknowledgements
interface PacketTimeStamp<T32khz, uint32_t> as
```

PacketTimeStamp32khz
interface PacketTimeStamp<TMilli, uint32_t> as
PacketTimeStampMilli
interface Receive[am_id_t]
interface Receive as Snoop[am_id_t]
interface SplitControl

Interface SplitControl is used by EasyCollectionC which is renamed. It uses interface SplitControl as RadioControl. Commands and events are used by the EasyCollection module which is explained later. SplitControl looks like:

```
command error_t start ()
//Start this component and all of its subcomponents
command error_t stop ()
//Start this component and all of its subcomponents
event void startDone (error_t error)
//Notify caller that the component has been started
//and is ready to receive other commands
event void stopDone (error_t error)
//Notify caller that the component has been stopped.
```

Most of the collection interfaces (RoutingControl, RootControl and Receive) are provided by the CollectionC module. The send interface is provided by CollectionSenderC which is a virtualized collection sender abstraction module following an extract of the signature of the CollectionC module (tos/lib/net/ctp/-CollectionC.nc) and CollectionSenderC (tos/lib/net/ctp/CollectionSenderC).

```
configuration CollectionC {
  provides {
    interface StdControl;
    interface Send[uint8_t client];
    interface Receive[collection_id_t id];
    interface Receive as Snoop[collection_id_t];
    interface Intercept[collection_id_t id];
    interface Packet;
    interface CollectionPacket;
    interface CtpPacket;
    interface CtpInfo;
    interface CtpCongestion;
    interface RootControl;
  }

generic configuration CollectionSenderC(collection_id_t
  collectid) {
  provides {
    interface Send;
    interface Packet;
  }
```

Note that the sender and receive interfaces require a collection_id_t to distinguish different possible collection trees. The CollectionC module provides some other interfaces in addition to the ones used in this example. As we explained previously, the CollectionC module generates a collection tree that will be used for the routing. These interfaces can be used to get information or modify the routing tree. For instance, if we want to obtain information about this tree we use the CtpInfo interface (see tos/lib/net/ctp/CtpInfo.nc) and if we want to indicate or query whether any node or sink is congested, we use the CtpCongestion interface (see tos/lib/net/ctp/CtpCongestion.nc).

EasyCollectionC uses the Timer, Leds and Boot interfaces and the following shows the interfaces:

```
toss/interfaces/Boot.nc:
interface Boot {
event void booted();
}


tos/interfaces/Leds.nc:
interface Leds {
/**
* Turn LED  on, off, or toggle its present state.
*/
async command void led0On();
async command void led0Off();
async command void led0Toggle();
async command void led1On();
async command void led1Off();
async command void led1Toggle();
async command void led2On();
async command void led2Off();
async command void led2Toggle();
/**
* Get/Set the current LED settings as a bitmask.
* Each bit corresponds to whether an LED is on;
* bit 0 is LED 0, bit 1 is LED 1, etc.
*/
async command uint8_t get();
async command void set(uint8_t val);
}


tos/lib/timer/Timer.nc:
interface Timer
{
// basic interface
command void startPeriodic( uint32_t dt );
command void startOneShot( uint32_t dt );
```

```
command void stop ();
event void fired ();
// extended interface omitted (all commands)
}
```

Regarding the interfaces for Boot, Leds and Timer, we can see that since EasyCollectionC uses those interfaces it must implement handlers for the Boot.booted() event, and the Timer.fired() event. The Leds interface signature does not include any events, so EasyCollectionC cannot implement any in order to call the Leds commands. Here, again, in EasyCollectionC's implementation of Boot.booted():, EasyCollectionC uses three instances of the TimerMilliC component, wired to the interfaces Timer0, Timer1 and Timer2. The Boot.booted() event handler starts each instance. The parameter to startPeriodic() specifies the period in milliseconds after which the timer will fire (millseconds are used because of the TMilli in the interface). Because the timer is started using the startPeriodic() command, the timer will be reset after firing such that fired() is triggered every n milliseconds. Invoking an interface command requires the call keyword, and invoking an interface event requires the signal keyword. EasyCollectionC does not provide any interfaces, so its code does not have any signal statements.

Finally, to compile this application we need the following makefile:

```
COMPONENT=EasyCollectionAppC
CFLAGS += -I$(TOSDIR)/lib/net \
          -I$(TOSDIR)/lib/net/le \
          -I$(TOSDIR)/lib/net/ctp
include $(MAKERULES)
```

Now we can install this application into several nodes (as discussed earlier, base station is a node whose ID is 1). This application demonstrates how all the packets generated in the nodes are collected in the base station.

7.2.4   Modified SPIN in TinyOS

In this section, a brief description of the M-SPIN routing protocol and its implementation details is presented. The detail of M-SPIN is discussed in Chapter 3. Distance discovery and the negotiation phase are briefly discussed first to explain the implementation of M-SPIN.

Distance discovery. As hop distance is measured from the sink, the sink initially broadcasts a startup packet in the network with type, nodeId and hop. Here type means type of message. The nodeId represents the ID of the sending node and hop represents hop distance from the sink node. The initial value of the hop is set to 1. When a sensor node receives the startup packet, it stores this hop value as its hop distance from the sink node in memory. After storing the value, the sensor node increases the hop value by 1 and then re-broadcasts the startup packet to its neighbor nodes with the modified hop value. It may also be possible for a sensor node to receive multiple startup packets from different intermediate nodes. Whenever a sensor node b receives startup packets from its neighbors a, it checks the hop distances and sets the distance to the minimum. This process is continued until all nodes in the network get the startup packets at least once within the distance discovery phase. After successful completion of this phase, negotiation will be started.

Negotiation. The source node sends an ADV message. Upon receiving an ADV message, each neighbor node verifies whether it has already received or requested the advertised data. The receiver node also verifies whether it is nearer to the sink node or not in comparison with the node that has sent the ADV message. If the hop distance of the receiving node (own_hop) is less than the hop distance received by it as part of the ADV message ($rcev\_hop$), i.e., $own\_hop < rcev\_hop$, then the receiving nodes send a REQ message to the sending node for current data. The sending node then sends the data to the requesting node using DATA message.

In order to generate network traffic, a simple application is required. The required components for the overall routing mechanism are divided into two modules: (1) application module and (2) routing module. For the M-SPIN routing protocol we define a simple nesC application named as MymhMsgM.nc. It is a component stored into apps/MymhMsg in the TinyOS tree. It uses a timer periodically for requesting data from the sensor node and sends the current data over the network. The application also contains a MymhMsg.h header file and MymhMsg.nc configuration. MymhMsgM.nc actually provides the implementation of the M-SPIN application. MymhMsg.nc is used to wire the MymhMsgM.nc module to other components that the M-SPIN application requires. Now let's look at the configuration, MymhMsg.nc; first:

```
includes MyMsg;
configuration MymhMsg{
}

implementation
{
        components Main, Temp, MymhMsgM, TimerC, GenericComm
        as Comm, QueuedSend, SimpleTime, MHSpinRouter as
        Router;

        Main.StdControl -> MymhMsgM.StdControl;
        Main.StdControl -> Router.StdControl;
        Main.StdControl -> Comm.Control;
        Main.StdControl -> QueuedSend.StdControl;
        Main.StdControl -> TimerC.StdControl;
        Main.StdControl -> SimpleTime.StdControl;
        MymhMsgM.ADC -> Temp.TempADC;
        MymhMsgM.Timer -> TimerC.Timer[unique(''Timer'')];
        MymhMsgM.Send -> Router.Send[AM_MYMSG];
        MymhMsgM.GlobalTime -> Router.GlobalTime;
        Router.ReceiveMsg[AM_MYMSG] -> Comm.ReceiveMsg
```

```
        [AM_MYMSG] ;
}
```

As discussed earlier, the configuration is implemented within the pair of curly brackets following the implementation key word. The components line specifies the set of components that this configuration references, in this case Main, Temp, MymhMsgM, TimerC, GenericComm, QueuedSend, SimpleTime, MHSpinRouter. The remainder of the implementation consists of connecting interfaces used by components to interfaces provided by others. MHSpinRouter is a configuration of the routing module which is used by the configuration of this application module.

    As mentioned earlier, nesC uses arrows to determine relationships between interfaces. The right arrow ($\rightarrow$) is used as *binds to*. The left side of the arrow binds an interface to an implementation on the right side. In other words, the component that uses an interface is on the left, and the component providing the interface is on the right. Therefore,

    MymhMsgM.Timer $\rightarrow$ TimerC.Timer[unique("Timer")];

is used to wire the timer interface used by MymhMsgM to the timer interface provided by TimerC. MymhMsgM.Timer on the left side of the arrow is referring to the interface called Timer (tos/interfaces/Timer.nc), while TimerC.Timer on the right side of the arrow is referring to the implementation of Timer (tos/lib/TimerC.nc).

    Now take a look at MymhMsgM.nc:

```
includes  MyMsg;
module  MymhMsgM
{
        provides
        {
                interface  StdControl;
        }
        uses
        {
                interface  ADC;
                interface  Send;
                interface  Timer;
                interface  GlobalTime;
        }
}

implementation
{
        norace  uint16_t  sensorReading;
    //  Holds  current  sensor  reading
        bool sendBusy; //  Holds  status  of  send
```

```
TOS_Msg mhMsg; // Message to use when sending
data static uint16_t seq;
```
...
```
}
```

The first part of the code states that this is a module called MymhMsgM
and declares the interfaces it provides and uses. The MymhMsgM module
provides the interface StdControl. This means that MymhMsgM implements
the StdControl interface. The MymhMsgM module also uses four interfaces:
ADC, Send, Timer and GlobalTime. This means that MymhMsgM may call
any command declared in the interfaces it uses and must also implement any
events declared in those interfaces. All the above components and configuration use the MyMsg.h file. This header file defines a packet structure named
as MyTestMsg.

```
enum
{
        TIMER_RATE = 30000,
        AM_MYMSG = 10,
};
```

```
typedef struct MyTestMsg
{
        uint16_t address;                      //source address
        uint32_t timestamp;
        uint16_t seqNo;
        uint16_t info;
} __attribute__ ((packed)) MyTestMsg;
```

The routing module is the second phase of this M-SPIN protocol. TinyOS
includes an optional multi-hop routing layer for applications which require
multi-hop functionality. The components used here were created or modified
during the implementation of the routing protocol. The original components
are parts of the standard TinyOS multi-hop routing library. Descriptions of
some essential components are given below.

MHEngineM is a modified version of MultiHopEngineM. It provides the
overall packet movement logic for multi-hop functionality. Using the Route-
Select interface, it determines the next-hop path and forwards the packet to
the parameterized SendMsg port. It also provides methods to send and receive
packets over the network. The MHEngineM component provides the Send interface to the higher layers in the networking stack and uses the ReceiveMsg and
SendMsg interfaces of lower networking layers to send and receive data. The
RouteSelect interface is used to communicate with the path selection modules.

```
        includes AM;
        includes MH;
```

```
#ifndef  MHOP_QUEUE_SIZE
#define  MHOP_QUEUE_SIZE  32
#endif

module  MHEngineM
{
      provides
      {
            interface  StdControl;
            interface  Receive[uint8_t  id];
            interface  Send[uint8_t  id];
            interface  Intercept[uint8_t  id];
      }
      uses
      {
            interface  ReceiveMsg[uint8_t  id];
            interface  SendMsg[uint8_t  id];
            interface  RouteSelect;
            interface  StdControl  as  SubControl;
            interface  StdControl  as  CommStdControl;
            interface  GlobalTime;
            interface  SpinLog;
            interface  Notify;
#ifdef  SIM
            interface  Timer  as  SimTimer;
#endif
            interface  Leds;
      }
}
```

MHSpinPSM is the path selection module for the M-SPIN protocol. It is responsible for providing the MHEngineM module with a route decision via the RouteSelect interface. When using the M-SPIN protocol, an image of the original data, i.e., ADV packets, are sent to every node within range by addressing all packets to the broadcast address. The base station does not advertise the packets but rather addresses packets to the UART address. For simplicity, it is assumed that *num_node*1 will always be the sink. Here *num_node* is the maximum number of nodes given for simulation. By choosing the next-hop address in this way, a valid route from source to destination will be available. All packets pass to the module via the RouteSelect interface.

```
includes  AM;
includes  MH;
includes  MyMsg;
module  MHSpinPSM
{
```

```
                provides
                {
                        interface  StdControl;
                        interface  RouteSelect;
                        interface  SpinLog;
                }
        }
```

MHSpinRouter configuration connects MHEngineM and MHSpinPSM with other necessary components. The configuration exports Receive, Send and Intercept and Snoop (as Intercept) ports to applications.

```
        includes MH;

        configuration  MHSpinRouter
    {
                provides
                {
                        interface  StdControl;
                        interface  Receive[uint8_t  id];
                        interface  Send[uint8_t  id];
                        interface  Intercept[uint8_t  id];
                        interface  GlobalTime;
                }
                uses
                {
                        interface  ReceiveMsg[uint8_t  id];
                }
    }

implementation
{
components MHEngineM, MHSpinPSM, GenericComm  as  Comm,
QueuedSend, TimerC, RandomLFSR, SimpleTime, TimeSyncM, LedsC,
MHSpinAdvM, MHSpinReqM, HelloSendM, HelloReceiveM;

  ...
}
```

The HelloSendM module is used for the distance discovery phase.

```
        includes AM;
        includes MH;
        module HelloSendM
        {
            ...
            ...
        }

        implementation
```

```
{
        ...
        typedef struct HopTable
        {
                uint16_t hop;   // variable to store hop_count
                value

        } HopTable;

        TOS_Msg routeMsg;  // variable to use to store route
        messages
        bool sendRouteBusy;  // indicates that send is busy
        uint16_t fwdCount;  // a counter variable
        HopTable ht;

        void hello_pkt(uint16_t cn)
        {
        HelloMsg *pHello = (HelloMsg *) &routeMsg.data[0];
        uint8_t length = sizeof(HelloMsg);
                // do not send if channel is in use
        if (sendRouteBusy == TRUE)
        {
                return;
        }
        pHello->type = Hello_TYPE;
        pHello->originNode = TOS_LOCAL_ADDRESS;
        pHello->counter = cn;
        fwdCount++;

        if (fwdCount>=MAX_HOP)
        {
                return;
        }
        if (call SendMsg.send(TOS_BCAST_ADDR,
            length, &routeMsg) == SUCCESS)
        {
                atomic
                {
                        call Leds.redToggle();
                }
                atomic sendRouteBusy = TRUE;
        }
        }
        ...
}
```

Again, in order to run the application, the user can modify the makefile and put it into the application folder. The makefile for the MymhMsgM component

```
SIM: Random seed is 578125
3: TimeSyncM — Node synchronised to global time
2: TimeSyncM — Node synchronised to global time
1: TimeSyncM — Node synchronised to global time
4: TimeSyncM — Node synchronised to global time
5: TimeSyncM — Node synchronised to global time
0: HelloSendM — Hello Packet sent from 0 node —> 1
3: HelloReceiveM — Hello Packet Received from node 0 — Counter 1
3: HelloSendM —updHop — 1
3: HelloSendM — Hello Packet sent from 3 Node —> 2
1: HelloReceiveM — Hello Packet Received from node 3 — Counter 2
1: HelloSendM —updHop — 2
1: HelloSendM — Hello Packet sent from 1 Node —> 3
0: HelloReceiveM — Hello Packet Received from node 3 — Counter 2
2: HelloReceiveM — Hello Packet Received from node 3 — Counter 2
2: HelloSendM —updHop — 2
2: HelloSendM — Hello Packet sent from 2 Node —> 3
4: HelloReceiveM — Hello Packet Received from node 2 — Counter 3
4: HelloSendM —updHop — 3
4: HelloSendM — Hello Packet sent from 4 Node —> 4
2: HelloReceiveM — Hello Packet Received from node 4 — Counter 4
5: HelloReceiveM — Hello Packet Received from Node 4 — Counter 4
5: HelloSendM —updHop — 4
5: HelloSendM — Hello Packet sent from 5 node —> 5|
4: HelloReceiveM — Hello Packet Received from node 5 — Counter 5
5: MymhMsgm — (ori,seq,data) = 16440(5,1,0)
5: MHSpinAdvM — ADV — originNode 5,Seq 1,hc 4
4: MHSpinReqM — REQ sent to node 5
3: MymhMsgM — (ori,seq,data) = 16598(3,1,0)
3: MHSpinAdvM — ADV — originNode 3,Seq 1,hc 1
1: ADV from better node — hc 1
2: ADV from better node — hc 1
0: MHSpinReqM — REQ sent to node 3
0: MymhMsgM — (ori,seq,data) = 20000(0,1,0)
0: MHSpinPSM — (ori,seq,data) = (0,1,0)
0: MHEngineM — Sending of packet successfull
```

FIGURE 7.3

Sample output in TOSSIM

is shown below. COMPONENT of the Makefile should be the name of the Configuration used. Here it is MymhMsg.

```
PLATFORMS=mica mica2 mica2dot micaz pc
COMPONENT=MymhMsg
PFLAGS= −I\%T/lib/SPIN3 −I\%T/lib/Queue −I\%T/lib/Broadcast
include ../Makerules
```

A sample output generated in TOSSIM is shown in Figure 7.3.

7.3   Contiki

The COOJA simulator of Contiki is described in this section followed by a discussion on Contiki installation and an overview of implementation of a simple application namely, a broadcast-example.

7.3.1   COOJA

The Contiki operating system Java, also known as COOJA, is a sensor network simulator for the Contiki OS [2]. Programming languages for the COOJA simulator include C and Java which offer quick and easy ways to develop new application modules and extend them later. The ability to compile and execute Contiki OS code in a simulator minimizes the hurdles a developer may encounter while actually running the code on a real platform. The Java native interface (JNI) establishes a well-defined and platform-independent interface between the C language of Contiki code and the pure Java part of the simulator. Native code can be used with Java in two distinct ways: as "native methods" in a running JVM and as the code that creates a JVM using the invocation API. Java native methods are declared in Java, implemented in C language, and loaded by the JVM as necessary. The reason for using JNI is to reuse libraries and APIs not implemented in Java. The main design goal of COOJA is extendibility. JNI is a good choice to support this feature which helps to provide interfaces and plugins. New plugins and interfaces can also easily be created and added to the simulator. Mainly, the plugins interact with a simulation and an interface interacts with a sensor node. However, COOJA is not a sensor node emulator. The hardware peripherals of a platform are not emulated; they are replaced by other simulated devices. An alternative approach would be to emulate all hardware of the sensor, which obviously gives more precise results but also would limit the simulator to a few hardware platforms.

The core graphical user interface (GUI) of COOJA is very simple and it is based on a desktop pane. Users can easily create, load or save simulations using menus and dialogs. Simulator parameters and settings are controlled by the user with the help of plugins which is a very useful way to interact with a simulation.

7.3.1.1   Interfaces

In COOJA simulation, each node is connected to a node type. Each node has its own memory and a number of interfaces. The memory consists of one or several segments, each with a start address and data. The interfaces simulate node peripherals by updating data in the memory. For instance, when the time changes, a clock interface should update the time variable to reflect that. That variable resides in the node memory of a specific node. Node type plays an important role in bridging the simulated node and core of Contiki OS. In COOJA, a node is initialized at startup by the core and creates the initial memory for it. After that all the processing is performed remotely through JNI. All nodes of the same type are linked to the same loaded core. The node type also performs variable name-to-address mapping. For example, the clock interface can change the core time variable by getting the variable address from the node type. Thus, the node type is responsible for linking the node to the core. All nodes of the same type share the same Contiki OS as there is one running Contiki OS for each node type. In the simulation platform, the Contiki code is compiled as a shared library. The core is implemented in C and the simulator is written in Java. Therefore, the communication between the Java part and the core is through JNI. The core is responsible for the inner workings of the node which needs Contiki OS code. On the other hand, simulator handles external tasks of a node, e.g., the current simulation time. There are four native functions to manage the simulation:

- An `initialization` function starts the process handler, networking and the pre-specified application processes.

- A `get memory` function is used to get a specified byte array from the current process memory.

- A `set memory` function sets or returns a specified byte array from the current process memory.

- An `absolute memory address` function returns an absolute address of a reference variable which is used for mapping between relative and absolute memory addresses. Each node type knows the relative addresses of all memory sections and variables in its core. When the simulator switches between different nodes, all of the node-relevant process memory has to be replaced. and a `tick` function is called during a node tick. In other words, when the simulation is running, all sensors act in turn and the `tick` function is called for each node. During a node tick, the Contiki OS function `process_run()` is called, which handles the event and polls a process. In order to tick a node, the memory of that node has to be set before invoking core functions.

The COOJA interfaces interact with the simulated nodes. Some interfaces are used to simulate the hardware devices and some act as virtual interfaces to node properties. For example, for node position, by customizing the position interface, a simulation with mobile nodes can be created. COOJA interfaces exist both in the core and in the simulator. Interfaces implemented in the simulator have full access to the node memory. On the other hand, interfaces implemented in the core can access Contiki functions through JNI. Simulator interfaces can be made active or passive. The only node property not handled through the interfaces is the node state. A node is active, sleeping or dead. The difference between a sleeping and dead node is that a sleeping node can be ticked but a dead node will never be ticked and can never leave the dead state. When a sleeping node is ticked, passive interfaces act on it and the tick is not delivered to the core. On the other hand, active interfaces can wake a sleeping node by triggering an external interrupt. An example of passive interface is the battery interface, which must act even though the node is sleeping as it still needs battery energy. A button interface is an example of an active interface. Button interface is active and wakes a sleeping node whenever the button is pressed. All simulation interfaces can also be observed. Any entity of the simulator can register as an observer to get notifications from the simulation interfaces. For example, a radio interface sends notification to its observers before sending actual data. Radio medium is a standard observer of radios. After receiving the notification, it gets ready to fetch the new data and decide which of the other radios should receive it. Each radio medium is listening to a number of registered radio transmitters. Every radio transmitter has its own position. The radio medium gets notifications whenever new data arrives at any of the radios and then decides which other radios should receive the data.

This observer-observable approach is interesting and enables dynamic interactions between different parts of the simulator.

7.3.1.2   Plugins

Plugins allow a user to interact with a simulation. The Contiki system provides services which the plugins can use. Plugins are registered at runtime, often at startup of the simulator with the core. The user then creates instances of the available registered plugins during simulations. Plugins depend on the services provided by the core and do not usually work by themselves. Conversely, the core operates independently of the plugins. Moreover, users can easily add and update plugins without making changes to the core. The plugins are implemented like a regular Java panel. Hence a user can easily create new graphical interfaces. The four plugin types are:

- *GUI plugin:* This type only needs a running GUI to be constructed and this is passed as an argument when a user initializes the plugin. Generally, this plugin is optional. However, users can access relevant information such as the current simulation as well as all simulated nodes through the GUI. This type of plugin only depends on the GUI, so it is not removed when the current simulation is removed.

- *Simulation plugin:* This type depends on a simulation. The current simulation is passed as an argument to the simulation plugin when it is created. The plugin is removed when the simulation is removed. When a new simulation is created, the simulation plugin can be automatically created. A very useful application of the simulation plugin is to display information about the current active simulation, for example, number of nodes, their types, positions and simulation status.

- *Mote plugin:* This plugin depends on a simulated node. Therefore, when a node is removed, the plugin is also removed. This plugin is useful to monitor node behaviour, for example, it can watch a certain node variable and stop the simulation based on the predefined value of the variable. An energy usage history plugin is a good example of this type which is very useful in monitoring the available battery energy of a node and displaying a graph over time using monitored data.

- *Dynamic plugin:* Any plugin type can be registered as a dynamic plugin. COOJA supports registering and un-registering plugins as per the requirements of the current simulation. By using the dynamic plugin, a simulation can register one or more plugins of its own. Radio medium is a good example of this type. A generic visual interface is not always enough since radio mediums can be very advanced. Therefore, using dynamic plugins, a radio medium can register one or several plugins of its own.

Some popular plugins that offer basic functionality are implemented in the basic version of COOJA. This includes viewing and moving nodes, monitoring radio traffic, monitoring node log outputs and controlling simulations.

7.3.2   Contiki Installation

The Berkeley Software Distribution organization and community of Contiki users have provided an easy method for installing and using the COOJA simulator. We will discuss the installation in a Windows 7 operating environment using an Intel Core^{TM} i3-2330M CPU (2.20 GHz, 3.00 GB RAM, 64-bit OK). The installation steps are:

1. Download InstantContiki from http://www.contiki-os.org/download. html. This will take some time because the file has an approximate size of 1.8 GB. The downloaded file is called *InstantContiki2.6.1.zip*. Unzip the downloaded file to a suitable drive (like C:/) that has more than 4 GB of free space. *InstantContiki2.6.1* containing files named *instant_Contiki_Ubuntu_12.04_32-bit-s001.vmdk*, *Instant_Contiki_Ubuntu_12.04_32-bit.vmx* and so on will be created.

2. Download and install VMWare Player from the http://www.vmware. com/go/downloadplayer/website. The download is free but registration is required. The installation may require a reboot of the computer.

3. Open VMWare Player and choose the "Open a Virtual Machine" option. See the first screenshot.

4. Select the *Instant_Contiki_Ubuntu_12.04_32-bit.vmx* file from the *InstantContiki2.6.1* unzip folder. Power on Contiki and wait for the virtual Ubuntu Linux to boot; see the second and third screenshots.

5. An InstantContiki login window will appear. Enter a *user* password to log in; see the fourth screenshot.

After InstantContiki is up and running, the COOJA simulation can be installed using the following steps:

1. Run the COOJA simulator

   user@instant-contiki:~$ cd contiki/tools/cooja

   user@instant-contiki:~/contiki/tools/cooja$ ant run

2. Create new simulation dialog by clicking *File>New simulation*.

3. Select a mote to emulate. The network window shows all the motes in the simulated network. Select *Sky mote...* to create an emulated *Tmote Sky mote* located in *Motes>Add motes...>Create new mote types...>Sky mote...*

4. Simulation will start and is controlled from the simulation control window using the Start, Pause, Stop and Reload buttons.

The detailed installation procedure is available at http://www.contiki-os.org/start/html.

7.3.3 Broadcast Example in Contiki

In this section, we discuss the simulation of an existing broadcast example. The file contains a simple Contiki application that randomly broadcasts a UDP packet to its neighbors.

1. After selecting *Sky mote* as described above choose a Contiki application. Go to the /home/user/contiki/examples/ipv6/simple-udp-rpl directory that contains a number of examples of simple UDP communications using IPv6. Choose the *broadcast-example.c* file.

2. Compiling the file will take some time. Next create a mote type.

3. Add motes from the newly created mote type to the simulation. You can also set the number of motes and their positioning.

4. After about 10 motes are added to the simulation in the network window, you can start the simulation. Printouts from the simulated motes appear on the mote output window. The network window shows continuing communication in the network. The timeline window indicates communication and radio events over time; the small gray lines represent radio awakenings by Contiki-MAC. Simulation may be paused by clicking the Pause button.

The application opens a UDP broadcast connection and sends one packet every second. Figure 7.4 shows a screenshot of this simulation.

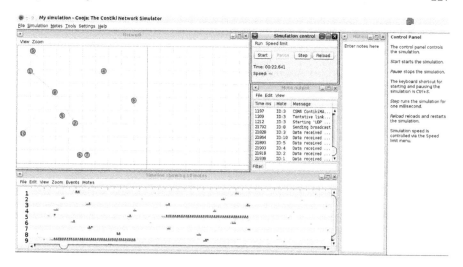

FIGURE 7.4 (SEE COLOR INSERT)
Screenshot of Contiki broadcast example

All Contiki programs must have processes; the declaration is:

PROCESS(example_program_process, "Example process");

Then, an event timer (etimer) is used to make the program send a packet every second as:

static struct etimer timer;

Next, the main program or process has to be implemented within the process thread function. The process is run whenever an event occurs, and the parameters *ev* and *data* are used to the event type and any data that may be passed along with the event. The basic structure of this program is given below.

```
#include "contiki.h"
#include "contiki−net.h"
PROCESS(example_program_process, "Example process");
static struct etimer timer;
PROCESS_THREAD(example_program_process, ev, data)
  {
  /* UDP connection is declared */
    static struct uip_udp_conn *c;
  /* process thread starts */
    PROCESS_BEGIN();
  /* create an UDP connection to port number 4321 */
    c = udp_broadcast_new(UIP_HTONS(4321), NULL);
    while(1)
```

```
{
/* Set a timer that wakes up once every second */
     etimer_set(&timer, CLOCK_SECOND);
     PROCESS_WAIT_EVENT_UNTIL(etimer_expired(&timer));

/* In order to send a UDP packet, the uIP TCP/IP
stack process (uIP works under the Hollywood principle:
"Do not call us, we will call you") */

     tcpip_poll_udp(c);
     PROCESS_WAIT_EVENT_UNTIL(ev == tcpip_event);
     uip_send("Hello", 5);
  }
  PROCESS_END();   //end of process
}
```

7.4   Castalia

Castalia is a simulator for WSNs and body area networks (BANs). It was developed at the National ICT Australia (NICTA) in 2006 [6]. In 2007, Castalia was made public as an open source project under an academic public license [5]. In most of the existing simulators for WSNs the designed models remain simplistic or unsuitable for short-range low-power communications where the impact to the result can be significant [4]. Castalia is popular for providing short-range low-power modelling for sensor networks. It also works for general networks with low-power embedded devices. It is based on the OMNeT++ platform and it can be used to check the validation of designed algorithms or protocols in realistic wireless channel and radio models. Specific application-based frameworks with different platform and architectures for wireless sensor networks such as hybrid wireless sensor networks, hierarchical wireless sensor networks, cognitive wireless sensor networks are also supported by Castalia. It can also be used to evaluate designed algorithms or protocols or frameworks in different types of WSNs such as terrestrial, underground, underwater, multimedia, mobile and mixed. Castalia provides parametric-based wireless channel and radio models.

Castalia provides the following features:

- *Channel model:* Simple and complex models include variation of path loss components when signal propagates from one node to another node. Castalia fully supports mobility of the nodes with clock drift facility. Interference and jamming are handled as received signal strength.

- *Radio model:* Castalia provides the following parameters in radio models to test protocols for low-power communication on different mediums:

  1. Multiple transmission power levels for sensors.
  2. Different power consumption and delay switching between sensors.
  3. Probability of reception based on signal-to-interference-plus-noise ratio (SINR), packet size, modulation type phase shift keying (PSK) or frequency shift keying (FSK) supported. Custom modulation is allowed by defining the SNR-BER curve.
  4. Realistic modelling of *received signal strength indication* (RSSI) and carrier sensing by which a sensor node determines when the amount of radio energy in the channel is below a certain threshold, at which time the sensor receives a clear-to-send (CTS) instruction to send data packets.

- *Process model:* Features are flexibility in physical process model, sensing noise, bias, and power consumption of individual nodes and calculation of average power consumption of networks. Some MAC and *routing protocols* are also available. Programmers can also design their own protocols.

Castalia is based on a message passing mechanism. A node collects the information from the lower physical layer and communicates with other nodes by passing the messages through a wireless channel. Figure 7.5 shows the connecting procedure in Castalia. The nodes are linked through the physical processes. This operation can be monitored by sampling the physical process in space and time to get information from the physical layer. A node is a composite which includes a sensor manager module, application module

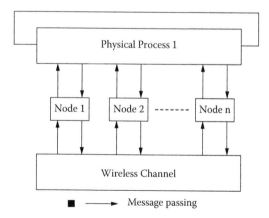

FIGURE 7.5
The modules and their connections in Castalia

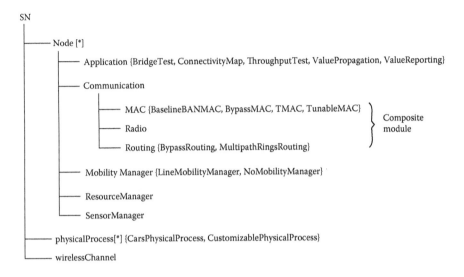

FIGURE 7.6
List of modules of Castalia

and communication module. Figure 7.6 shows the list of modules. To define a module in Castalia, we need to define a module name, module parameters, module interface, and possible submodule structure for composite module. OMNeT++ NED language is used for programming. In Castalia, every module corresponds to a directory which always contains a *.ned file that defines the module. For a composite module, there are subdirectories to define the submodules.

7.4.1 Modules under Castalia

Modules are discussed in this section. A list of modules of Castalia is shown in Figure 7.6. Descriptions of modules are as follows:

Mobility Module: This module controls the movement of sensor nodes and provides current positions of mobile sensors to the wireless channel periodically. To control the movements, locations of sensor nodes are divided into cells. If a node changes its position from one cell to another cell, the mobility module passes the current position of that node as a message to the other modules. Note that every module can be accessed from every other module by calling functions. There are two existing mobility managers in Castalia: lineMobilityManager and noMobilityManager.

- In lineMobilityManager, nodes move horizontally, vertically or diagonally. A programmer can program the movements as per requirements. Speeds and positions (X-coordinate, Y-coordinate and Z-coordinate) of sensor nodes can also be defined by the programmer into the simulation time in *.ini file.

- In noMobilityManager, no movements of nodes are allowed. That means nodes are assumed static.

Sensor Module: In the real world, many types of physical or environmental changes (humidity, temperature, light, etc.) need to be monitored by WSNs. In Castalia, the sensor module provides knowledge about sensing device capability which can be mapped to real world applications. Sensor nodes can sense multiple physical changes with the help of sensor modules. However, the default value of a number of physical changes is one; that is, only one physical change can be monitored by a sensor at any time. The default sample rate is one and the default sensor type senses temperature.

Resource Module: This module tracks information of each node during simulation such as estimated energy consumption of nodes, lifetime and clock drift estimation during channel sensing. Initial default values to start a simulation are as follows: initial power is 18720 joules, periodic power consumption of node is 6 mW, CPU clock drift is 0.00003 ms, and interval for energy calculation is 1000 msec.

Application Module: This module references the resource module, mobility module and radio module to access their public methods directly. There are three submodules to check the validation of each module: the throughput test, value propagation test and value reporting test.

Communication Module: There are three submodules in the communication module: MAC, radio and routing. An overview of a MAC submodule is as follows:

- TMAC and BMAC are available.

- The routing submodule of Castalia allows multipath ring routing and no routing (single-hop message transmission). During multipath ring routing, the sink node generates a signal. If the surrounding nodes receive the signal tracked to level 1, they send an acknowledgment to the sink. The process continues until every node has a level 1 transmitted or received signal. To send a packet to the sink, a node broadcasts the packet to the same-level nodes and waits until the packet reaches to the sink.

- The radio module provides a facility to capture many features of low-power radio nodes and channels. Some features are listed below:

  1. State: transmit, receive listen or sleep.
  2. Delays: Time delay to switch one state to other.
  3. Transmission power level: Provides multiple power levels of transmission.
  4. Power consumption: Power consumption of different states.
  5. Modes of operation: Datarate, bandwidth, noise-floor, modulation and so on.
  6. Multiple modulation: FSK, PSK and others.

7. RSSI: Continuously calculates the received signal strength indicator.

To initialize the above features, three radios based on IEEE 802.15 are CC1000.txt, CC2420.txt and BANRadio.txt, located in Castalia>Simulations> Parameters>Radio.

7.4.2   Castalia Installation

Castalia 3.2 requires OMNeT Versions 4.1 or 4.2. Note that older versions of Castalia may require older versions of OMNeT. First, users have to install OMNeT++, then Castalia. OMNeT++ can be installed in Linux, Windows or Cygwin under Windows. We prefer Ubuntu 13.04 for Castalia installation. However, Redhat, Suse, Mint can also be used. The installation procedure in Ubuntu 13.04 is as follows:

OMNeT++ installation: Run the following commands:

1. $ sudo apt-get update
2. $ sudo apt-get install build-essential gcc g++ bison flex perl tcl-dev tk-dev blt libxml2-dev zlib1g-dev default-jre doxygen graphviz libwebkitgtk-1.0-0 openmpi-bin libopenmpi-dev libpcap-dev

    At the confirmation questions (Do you want to continue? [Y/N], answer Y). Make sure the Tcl/Tk shell is properly installed.
3. $ wget http://www.omnetpp.org/omnetpp/doc_download/2217-omnet-41-source–ide-tgz

    This is a large file (148 MB) so it might take time to download. It also can be downloaded from http://www.omnetpp.org/omnetpp/cat_view/17-downloads/1-omnet-releases. Assume the downloaded file is omnetpp-4.3.1-src.gz.
4. To untar the source file type

    $ tar xvfz omnetpp-4.3.1-src.gz

    A directory named omnetpp-4.3.1 will be created.
5. $ cd omnetpp-4.3.1
6. To set environment variables, type

    $ export PATH=$PATH:~/omnetpp-4.3.1/bin
    $ export LD_LIBRARY_PATH=~/omnetpp-4.3.1/lib

    Also add the above two export commands at the end of the .bash_profile file.
7. $ ./configure

    Do not use *NO_TCL*=1./ to configure because the configure script needs a graphical environment to test existing protocols (such as ALOHA) under OMNeT++/

8. Compile OMNeT++

    $ make

    Now the sample simulations should run correctly. For example, the dyna simulation is started by entering the following commands:

    $ cd samples/dyna
    $ ./dyna

9. To start the IDE, type

    $ omnetpp
    $ make install-menu-item
    $ make install-desktop-icon

Details of installation and overview of OMNeT++ are available in Reference [9].

Castalia Installation: Download the source code from [1]. The downloaded file will be Castalia-3.0.tar.gz or castalia.zip. To install Castalia, type the following commands:

1. Untar and unzip the source code:

    $ tar -xvzf Castalia-3.0.tar.gz

    A new directory named Castalia-3.0/ will be created

2. To build Castalia, type

    $ cd Castalia-3.0/
    $ ./makemake

    If the access to the script is refused, type $ chmod u+x makemake and try again. This automatically generates a makefile.

    A new directory named Castalia-3.0/ will be created.

3. To build Castalia, type

    $ make

    Check that the CastaliaBin soft link is created in Castalia-3.0/.

Castalia should have been built successfully under OMNeT++ platform. Detailed descriptions are available [3]. If a new module with *.cc, *.h, *.ned and *.msg files is created, make sure that the files and directory structures follow the module structure, then rebuild Castalia by typing following commands:

    $ make clean
    $ ./makemake
    $ make

TABLE 7.1

TMAC protocol parameters (TMAC.ned)

Parameter	Description
maxTxRetries = default (2)	maximum number of re-transmissions
allowSinkSync = default (true)	starts node synchronization
useFrts = default (false)	default FRTS facility not implemented
useRtsCts = default (true)	allows RTS/CTS handshake to be enabled or disabled
disableTAextension = default (false)	effectively creates SMAC protocol
resyncTime = default (6)	time in seconds for resending SYNC message
contentionPeriod = default (10)	10 ms
conservativeTA = default (true)	number of stay-awake milliseconds after channel is sensed
listenTimeout = default (15)	timeout of activation event (milliseconds)
waitTimeout = default (5)	expected time for reply (milliseconds)
frameTime = default (610)	frame time (milliseconds)
collisionResolution = default (0)	low collision avoidance (immediate retry)

7.4.3   TMAC in Castalia

This section describes the implementation details of the TMAC protocol [12] (discussed in Chapter 4) in Castalia. The four main components are TMAC.cc (for the TMAC algorithm), TMAC.h (the header file of the TMAC algorithm), TMAC.ned (the TMAC network structure) and TMacPacket.msg (the TMAC message structure).

TMAC.ned inherits the iMac.ned file which is a root component of the network structure used in the MAC protocol. In TMAC.ned, the sizes of the MAC packets are initialized and their descriptions are defined in TMAC's message component. The frame size, data frame overhead, MAC buffer size and TMAC protocols are also initialized in the TMAC.ned file. Table 7.1 lists the TMAC protocol parameters.

TMacPacket.msg Different types of TMAC packets are defined and extend the super class MacPacket. Types of MAC packets are SYNC as 1, Request to send (RTS) as 2, Clear to Send (CTS) as 3, DS as 4, FRTS as 5, DATA as 6 and ACK as 7. The MAC packet contains NAV vector and sequence number, value, and ID fields.

TMAC.h is the header file of TMAC.cc. It initializes the variables used in TMAC.cc such as MAC states (sleep, active, active but silent, RTS/CTS sense, ACK/DATA sense, etc.), sets timer for TMAC like SYNC create, SYNC send, DATA start, carrier sense and transmission time out.

TMAC.cc There are different methods in TMAC.cc to design TMAC protocols. The methods use different parameters to work properly from other modules like radios and wireless channels. The methods and their functionalities are listed in Table 7.2.

Now we will discuss how to simulate TMAC protocol from different application's test module views. Test modules (radioTest, valuePropagation and valueReporting) are located in Castalia>Simulations. All test modules contain a omnetpp.ini file. By taking help from this initialization file, simulation will start. In a typical omnetpp.ini file, simulation time limit, size of deployment field, number of nodes, deployment type, radio parameter file (default is CC2420.txt), MAC frame size and physical layer frame size are initialized.

We will show how to simulate radioTest of TMAC to get transmission delay of packets. There are three nodes to test throughput. Add *SN.node[*].Communication.MACProtocolName = "TMAC"* statement in the omnetpp.ini file. To simulate radio test with default interference configuration, write the following commands in terminal:

~/Castalia/Simulations/radioTest$ ../../bin/Castalia -c General
        Running configuration 1/1
~/Castalia/Simulations/radioTest$ ls
        250814-154911.txt   Castalia-Trace.txt   omnetpp.ini

It generates two files: an output text file (250814-154911.txt) and a trace file) Castalia-Trace.text) into the radioTest module. Note that the format of the text file is DDMMYY-HHMMSS.txt. To show general output, input the following command:

~/Castalia/Simulations/radioTest$ ../../bin/CastaliaResults -i 250814-154911.txt

We can also simulate combinations of two or more configurations as shown below:

~/Castalia/Simulations/radioTest$ ../../bin/Castalia -c [InterferenceTest1, InterferenceTest2] varyInterferenceModel
        Running configuration 1/2
        Running configuration 1/2
~/Castalia/Simulations/radioTest$ ls
        250814-155654.txt   Castalia-Trace.txt   omnetpp.ini

As shown in Figure 7.7, each output has $N \times M$ dimensions. N is the number of modules that produced this output and M is the number of indices of an output from a single module. For example, $M = 2$ if there are two senders.

Following is the list of commands to access these modules:

TABLE 7.2
Methods of TMAC protocol (TMAC.cc)

Method	Description
void startup()	initializes local parameters with MAC parameters from other components (*.h, *.msg, *.ned), sets MAC state, gets schedule from other sensors, creates SYNC packet and sends an interrupt signal if SYNC packet is ready
timerFiredCallback()	if timer fires, sets it for different types of packets to send
fromNetworkLayer()	creates a MAC frame from received packet by decapsulating and inserting retrieved information into MAC buffer
fromRadioLayer()	handles a MAC frame received from physical layer and checks whether received packet contains TMAC frame type, i.e., sender ID, destination ID and nav vector value
handleRadioControlMessage()	controls radio interrupt messages; if it senses interruption in channel, it generates busy signal, receives interrupt message and passes it to network layer
resetDefaultState()	resets MAC state: (1) goes to sleep if still active and timeout is expired; (2) if receiving schedule table, transmission can be handled after random waiting time; (3) switches state to active silent if no schedule table is in process and current state is listening
setMacState()	traces and finds reasons to change MAC states
createPrimarySchedule()	creates synchronization sechdule
sendDataPacket()	if MAC buffer is not empty, sends first data front from buffer to radio layer
carrierIsClear()	handles carrier clear message from radio modules; to send RTS, creates new RTS packet and sends it to radio layer by switching radio state to transmission and so on for DATA, SYNC packets
carrierIsBusy()	if channel is busy, extends active period and sets MAC state to active silent unless receiving other packets
updateScheduleTable()	updates synchronization schedule with a given value of wakeup time, schedule ID and sensor node ID
performCarrierSense()	sets timer to perform carrier sense
extendActivePeriod()	extends active period to ensure that remaining active time is greater than or equal to listening time out
checkTxBuffer()	checks whether transmission buffer is empty and updates number of re-transmissions
popTxBuffer()	fetches first packet of MAC from MAC buffer

Latency of packets: ~/Castalia/Simulations/radioTest$ ../../bin/CastaliaResults -i 250814-154911.txt -s application

Received packets: ~/Castalia/Simulations/radioTest$ ../../bin/CastaliaResults -i 250814-154911.txt -s RX

Transmitted packets: ~/Castalia/Simulations/radioTest$ ../../bin/CastaliaResults -i 250814-154911.txt -s TX

Module	Output	Dimensions
Application	Application level latency, in ms	1×2 (11)
	Packets received per node	1×2
Communication. Radio	RX pkt breakdown	3×1 (6)
	TXed pkts	2×1
ResourceManager	Consumed Energy	3×1

FIGURE 7.7
Different modules in Castalia

Energy consumption: ~/Castalia/Simulations/radioTest$ ../../bin/CastaliaResults -i 250814-154911.txt -s energy

Castalia also provides a facility to plot a graphical representation of outputs by taking the values from 250814-154911.txt. Write the following command to see the output graph:

~/Castalia/Simulations/radioTest$ ../../bin/CastaliaResults -i 250814-154911.txt -s application | ../../bin/CastaliaPlot -o latencymac.eps - -greyscale - -style=histogram - -yrange=0:200 - -xtitle="delay" - -ytitle="no. of packets" - -xrotate=45 - -hist-box=0.5

Figure 7.8 shows the graphical output generated from the above command. Other graph parameters are listed in Table 7.3.

Note that CastaliaResults automatically sets a time range (in milliseconds) between 0 to 200 in 11 groups for calculating application level latency as shown in Figure 7.7. Various interference test modules are considered here. For example, in Figure 7.8, node 0 receives more than 300 packets when considering InterferenceTest1 during simulation and each packet has a delay between 60 ms and 80 ms.

TABLE 7.3
Parameter used in graph display

Parameter	Description
grayscale	use greyscale colors only; default is color
style	plot style to be used, i.e., linespoints, histogram and stacked
yrange	set range of y-axis
xtitle	set title of x-axis
ytitle	set title of y-axis
xrotate	set x-axis rotation, in degrees
hist-box	set width of histogram columns

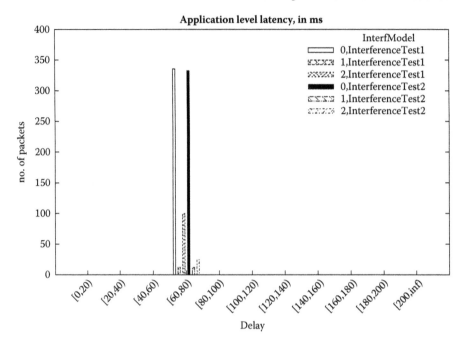

FIGURE 7.8
Latency output

7.5   NS-3

Network simulator is an important tool used to validate proposed algorithms. Researchers use it to check algorithms in various layers in networks like routing protocols, applications, mobility models, modifications in MAC layer and other events. NS-3 is a popular network simulator. It is free, open source and relies on C++ for the implementation of the simulation models. It provides advanced simulation environments for testing and debugging. The main program has to be written in C++ and is linked with static and dynamic libraries. Users can create test applications to define topology, number of nodes and routing protocols. This test script can be written in Python or in C++. However, the core parts of NS-3 and other existing algorithms are implemented in C++. It is very easy to add new protocols without much modification. Many popular algorithms are already implemented in NS-3. Hence, researchers can compare their proposed schemes with these existing algorithms. NS-3 also provides a stand-alone graphical tool called NetAnim to display the network topology and show the packet flow between nodes. NS-3 is supported on Linux $x86$ and $x86\_64$, FreeBSD $x86$ and $x86\_64$, Mac OS X ppc and $x86$. The detailed steps of downloading and building NS-3 and running the first script are available

TABLE 7.4
Required packages for NS-3

Package
gcc
gcc-c++
python
python-devel
bzr
gsl
gsl-devel
gtk2
gtk2-devel
gdb
valgrind
doxygen
graphviz
ImageMagick

[8] [7]. This section discusses how to install NS-3 on a Linux machine (Fedora or Ubuntu) and test existing sample programs. It also describes how to add a new mobility model and a corresponding sample program. Some software must be installed in the machine prior to installing NS-3. Table 7.4 shows the list of required packages.

The above packages can be installed by connecting a Linux machine to the Internet. For example, to install gcc we need to run following command (the user must have root permission to execute these commands):

- $ yum install gcc (this is for Fedora)

- $ apt-get install gcc (this is for Ubuntu)

7.5.1 Installation

NS-3 has lot of dependencies on other tools in Linux systems. All these packages can be installed as a single package and it is recommended to use the all-in-one package that has different folders as shown in Figure 7.9. Among these folders, NS-3.21 is the NS-3 main directory. Figure 7.10 shows directory structure of NS-3.21.

```
bake build.py constants.py constants·pyc netanim-3.105 ns-3.21 pybindgen-0.17.0.876 README util.py util.pyc
```

FIGURE 7.9
NS-3 all-in package

AUTHORS	different.pcap	doc	Makefile	scratch	testpy.supp	utils	waf	wutils.py
bindings	DlMacStats.txt	examples	ns3	src	UlMacStats.txt	utils.py	waf.bat	wutils.pyc
build	DlPdcpStats.txt	gtmots-animation.xml	README	test.py	UlPdcpStats.txt	utils.pyC	waf-tools	
CHANGES.html	DlRlcStats.txt	LICENSE	Release_NOTES	testpy-output	UlRlcStats.txt	VERSION	wscript	

FIGURE 7.10
Directory structure of NS-3.21

Follow the steps mentioned below to install NS-3 in a Linux system. We use ns-allinone-3.21 as an example. However, other versions could also be used and steps would be similar.

- Download NS-3 from

    https://www.nsnam.org/release/ns-allinone-3.21.tar.bz2

- Decompress the file using the following command

    tar -jxf ns-allinone-3.21.tar.bz2

- Change directory

    cd ns-allinone-3.21

- Compile NS-3 by

    ./build.py –enable-examples –enable-tests

- Change directory

    cd ns-3.21

- Test whether installation is successful by

    ./test.py -c core

- Run

    ./waf –run hello-simulator

If we get a message as shown in Figure 7.11, it confirms that NS-3 has been installed successfully.

```
'build' finished successfully (12.650s)
Hello Simulator
```

FIGURE 7.11
Output of hello simulator

7.5.2 Mobility Model in NS-3

Mobility models are very important to simulate mobile sensor networks. Here we discuss the method of adding a mobility model in NS-3. Existing supported mobility models can be found in ns-allinone-3.21/ns-3.21/src/mobility/model directory.

A random way-point (RWP) mobility model is one of the most widely used. Design of a mobility model can vary from one to another but certain steps are required to add a new mobility model (e.g., a modified RWP mobility model).

1. Write modified-random-walk-2d-mobility-model.cc

2. Write modified-random-walk-2d-mobility-model.h

3. Add entry for modified-random-walk-2d-mobility-model.cc in ns-3.21/src/mobility/wscript as shown in Figure 7.12.

4. Add entry for modified-random-walk-2d-mobility-model.h in ns-3.21/src/mobility/wscript as shown in Figure 7.13.

5. Compile by using the command ./waf from the NS-3.21 directory.

```
mobility.source = [
 'model/box.cc',
 'model/constant-acceleration-mobility-model.cc',
 'model/constant-position-mobility-model.cc',
 'model/constant-velocity-helper.cc',
 'model/constant-velocity-mobility-model.cc',
 'model/gauss-markov-mobility-model.cc',
 'model/hierarchical-mobility-model.cc',
 'model/mobility-model.cc',
 'model/position-allocator.cc',
 'model/random-direction-2d-mobility-model.cc',
 'model/random-walk-2d-mobility-model.cc',
 'model/random-waypoint-mobility-model.cc',
 'model/rectangle.cc',
 'model/steady-state-random-waypoint-mobility-model.cc',
 'model/waypoint.cc',
 'model/waypoint-mobility-model.cc',
 'model/sample-mobility-model.cc',
 'helper/mobility-helper.cc',
 'helper/ns2-mobility-helper.cc',
 ]
```

FIGURE 7.12 (SEE COLOR INSERT)
Entry of source file in wscript

In general, mobility models should define functions like GetTypeId(), DoInitialize(), DoGetPosition() and DoSetPosition(). However, detailed design of mobility models is beyond the scope of this book. The declaration of the mobility class in modified-random-walk-2d-mobility-model.h is as follows:

```
headers.source = [
  'model/box.h',
  'model/constant-acceleration-mobility-model.h',
  'model/constant-position-mobility-model.h',
  'model/constant-velocity-helper.h',
  'model/constant-velocity-mobility-model.h',
  'model/gauss-markov-mobility-model.h',
  'model/hierarchical-mobility-model.h',
  'model/mobility-model.h',
  'model/position-allocator.h',
  'model/rectangle.h',
  'model/random-direction-2d-mobility-model.h',
  'model/random-walk-2d-mobility-model.h',
  'model/random-waypoint-mobility-model.h',
  'model/steady-state-random-waypoint-mobility-model.h',
  'model/waypoint.h',
  'model/waypoint-mobility-model.h',
  'model/sample-mobility-model.h',
  'helper/mobility-helper.h',
  'helper/ns2-mobility-helper.h',
  ]
```

FIGURE 7.13 (SEE COLOR INSERT)
Entry of header file in wscript

```
class ModRandomWalk2dMobilityModel : public MobilityModel
{
public:
  static TypeId GetTypeId (void);
  enum Mode  {
    MODE_DISTANCE,
    MODE_TIME
  };

private:
  void Rebound (Time timeLeft);
  void DoWalk (Time timeLeft);
  void DoInitializePrivate (void);
  virtual void DoDispose (void);
  virtual void DoInitialize (void);
  virtual Vector DoGetPosition (void) const;
  virtual void DoSetPosition (const Vector &position);
  virtual Vector DoGetVelocity (void) const;
  virtual int64_t DoAssignStreams (int64_t);

  ConstantVelocityHelper m_helper;
  EventId m_event;
  enum Mode m_mode;
  double m_modeDistance;
```

```
    Time  m_modeTime;
    Ptr<RandomVariableStream> m_speed;
    Ptr<RandomVariableStream> m_direction;
    Rectangle m_bounds;
    bool m_constspeeden;
    double m_constspeed;
};
```

Each node usually has random speed in the random walk mobility model. The modified model maintains either active random speed or constant speed at one time. We introduce two variables to set speed: m_constspeeden defines whether constant or random speed is enabled; m_constspeed sets constant speed. Note that names for variables start with the m_prefix. A mobility model should include definitions of functions like ike GetTypeId(), DoInitialize() and DoGetPosition() and DoSetPosition(). Two attributes are added to the GetTypeId() example in modified-random-walk-2d-mobility-model.h; one is Boolean and the other is of double type as shown below:

```
TypeId
ModRandomWalk2dMobilityModel :: GetTypeId (void)
{
  static TypeId tid = TypeId ("ns3 :: ModRandomWalk2dMobilityModel")
   . SetParent<MobilityModel> ()
   . SetGroupName ("Mobility")
   . AddConstructor<ModRandomWalk2dMobilityModel> ()
   . AddAttribute ("Bounds",
                   "Bounds of the area to cruise.",
                   RectangleValue
                   Rectangle (0.0, 100.0, 0.0, 100.0)),
                   MakeRectangleAccessor
                   (&ModRandomWalk2dMobilityModel :: m_bounds),
                   MakeRectangleChecker ())
   . AddAttribute ("Time",
                   "Change current direction and speed after
                   moving for this delay.",
                   TimeValue (Seconds (1.0)),
                   MakeTimeAccessor
                   &ModRandomWalk2dMobilityModel :: m_modeTime),
                   MakeTimeChecker ())
   . AddAttribute ("Distance",
                   "Change current direction and speed after
                   moving for this distance.",
                   DoubleValue (1.0),
                   MakeDoubleAccessor
                   (&ModRandomWalk2dMobilityModel :: m_modeDistance),
                   MakeDoubleChecker<double> ())
   . AddAttribute ("Mode",
                   "The mode indicates the condition used to "
                   "change the current speed and direction",
                   EnumValue
                   (ModRandomWalk2dMobilityModel :: MODE_DISTANCE),
                   MakeEnumAccessor
                   (&ModRandomWalk2dMobilityModel :: m_mode),
                   MakeEnumChecker
```

```
                    ModRandomWalk2dMobilityModel :: MODE_DISTANCE,
                    "Distance",
                    ModRandomWalk2dMobilityModel :: MODE_TIME, "Time"))
  .AddAttribute ("Direction",
                    "A random variable used to pick the direction
                    (gradients).",
                    StringValue
                    ("ns3::UniformRandomVariable[Min=0.0|
                    Max=6.283184]"),
                    MakePointerAccessor
                    (&ModRandomWalk2dMobilityModel :: m_direction),
                    MakePointerChecker<RandomVariableStream> ())
  .AddAttribute ("Speed",
                    "A random variable is used to pick the speed
                    (m/s).",
                    StringValue
                    ("ns3::UniformRandomVariable[Min=2.0|Max=4.0]"),
                    MakePointerAccessor
                    (&ModRandomWalk2dMobilityModel :: m_speed),
                    MakePointerChecker<RandomVariableStream> ())
  .AddAttribute ("ConstSpeedEnable",
                    "If true, node moves in random direction
                    with constant velocity",
                    BooleanValue (false),
                    MakeBooleanAccessor
                    (&ModRandomWalk2dMobilityModel :: m_constspeeden),
                    MakeBooleanChecker ())
  .AddAttribute ("ConstSpeed",
                    "A double variable used to pick the speed (m/s).",
                    DoubleValue (4.0),
                    MakeDoubleAccessor
                    (&ModRandomWalk2dMobilityModel :: m_constspeed),
                    MakeDoubleChecker<double> ());
  return tid;
}
```

DoInitialize() could be considered the starting function. It calls DoInitializePrivate() which in turn calls DoWalk(). The control flow and an example of using those variables are shown in following code.

```
void ModRandomWalk2dMobilityModel :: DoInitialize (void)
{
  DoInitializePrivate ();
  MobilityModel :: DoInitialize ();
}

void ModRandomWalk2dMobilityModel :: DoInitializePrivate (void)
{
  m_helper.Update ();
  double speed = 0.0;
  double direction = m_direction->GetValue ();
  Vector vector (std::cos (direction) * speed,
                    std::sin (direction) * speed,
                    0.0);
  m_helper.SetVelocity (vector);
  m_helper.Unpause ();
```

```
    if  (m_constspeeden  ==  true )
      speed  =  m_constspeed ;
    else
      speed  =  m_speed−>GetValue  ();

    Time  delayLeft ;
    if  (m_mode  ==  ModRandomWalk2dMobilityModel :: MODE_TIME)
      {
        delayLeft  =  m_modeTime ;
      }
    else
      {
        delayLeft  =  Seconds  (m_modeDistance  /  speed );
      }
    DoWalk  (delayLeft );
}

void  ModRandomWalk2dMobilityModel :: DoWalk  (Time  delayLeft )
{
    Vector  position  =  m_helper . GetCurrentPosition  ();
    Vector  speed  =  m_helper . GetVelocity  ();
    Vector  nextPosition  =  position ;
    nextPosition . x  +=  speed . x  ∗  delayLeft . GetSeconds  ();
    nextPosition . y  +=  speed . y  ∗  delayLeft . GetSeconds  ();
    m_event . Cancel  ();
    if  (m_bounds . IsInside  (nextPosition ))
      {
        m_event  =  Simulator :: Schedule  (delayLeft ,
        &ModRandomWalk2dMobilityModel :: DoInitializePrivate ,  this );
      }
    else
      {
        nextPosition  =  m_bounds . CalculateIntersection  (position ,
        speed );
        Time  delay  =  Seconds  ((nextPosition . x  −  position . x)  /
        speed . x);
        m_event  =  Simulator :: Schedule  (delay ,
        &ModRandomWalk2dMobilityModel :: Rebound ,  this ,
                                          delayLeft  −  delay );
      }
    NotifyCourseChange  ();
}

void  ModRandomWalk2dMobilityModel :: Rebound  (Time  delayLeft )
{
    m_helper . UpdateWithBounds  (m_bounds );
    Vector  position  =  m_helper . GetCurrentPosition  ();
    Vector  speed  =  m_helper . GetVelocity  ();
    switch  (m_bounds . GetClosestSide  (position ))
      {
      case  Rectangle :: RIGHT:
      case  Rectangle :: LEFT:
        speed . x  =  −speed . x;
        break ;
      case  Rectangle :: TOP:
      case  Rectangle :: BOTTOM:
        speed . y  =  −speed . y;
```

```
      break;
    }
  m_helper.SetVelocity (speed);
  m_helper.Unpause ();
  DoWalk (delayLeft);
}
```

Rebound() is called by the scheduler as shown in below.

```
nextPosition = m_bounds.CalculateIntersection (position, speed);
Time delay = Seconds ((nextPosition.x - position.x)/speed.x);
m_event =Simulator::Schedule (delay,
            &ModRandomWalk2dMobilityModel::Rebound,
            this, delayLeft - delay);
```

Now we have to create an application which will use this newly added mobility model. We could take existing sample application programs as references, for example, the main-random-walk.cc located in ns-3.21/src/mobility/example directory. We need to use the mobility model shown below that also shows how to pass variables from the application program to the mobility model.

```
int main (int argc, char *argv[])
{
  bool csEnable = false;
  std::string animFile = "sample-animation.xml";
  // Name of animation file

  Config::SetDefault ("ns3::ModRandomWalk2dMobilityModel::Mode",
    StringValue ("Time"));
  Config::SetDefault ("ns3::ModRandomWalk2dMobilityModel::Time",
    StringValue ("2s"));
  Config::SetDefault ("ns3::ModRandomWalk2dMobilityModel::Speed",
    StringValue ("ns3::ConstantRandomVariable[Constant=1.0]"));
  Config::SetDefault ("ns3::ModRandomWalk2dMobilityModel::Bounds",
    StringValue ("0|200|0|200"));

  CommandLine cmd;
  cmd.AddValue ("csEnable", "Enable Constant Speed", csEnable);
  cmd.Parse (argc, argv);

  NodeContainer c;
  c.Create (1);

  MobilityHelper mobility;
  mobility.SetPositionAllocator ("ns3::RandomDiscPositionAllocator",
            "X", StringValue ("100.0"),
            "Y", StringValue ("100.0"),
            "Rho", StringValue ("ns3::UniformRandomVariable
            [Min=0|Max=30]"));
  mobility.SetMobilityModel ("ns3::\hl{ModRandomWalk2dMobilityModel}",
            "Mode", StringValue ("Time"),
            "Time", StringValue ("2s"),
            "Speed", StringValue ("ns3::ConstantRandomVariable
            [Constant=1.0]"),
            "ConstSpeedEnable", BooleanValue (csEnable),
            "Bounds", StringValue ("0|200|0|200"));
  mobility.InstallAll ();
```

```
Config :: Connect ("/ NodeList /*/ $ns3 :: MobilityModel / CourseChange" ,
              MakeCallback (&CourseChange ));

Simulator :: Stop (Seconds (100.0));

AnimationInterface anim (animFile);
Simulator :: Run ();

Simulator :: Destroy ();
return 0;
}
```

Here we specify the speed of each node as a random value; speed is the only variable of the sample mobility model. Most users keep their application programs in the scratch directory. Assume that our sample program name is main-sample.cc and we need to execute following command to run the application,

./waf run scratch/main-sample -csEnable=1

We can also set multiple parameters in the program. For example, if we want to set number of nodes as a parameter, we need to execute following command:

./waf –run "scratch/main-sample –numNodes=5"

For this, the sample application program should contain

cmd.AddValue ("numNodes", "number of nodes", numNodes);

NetAnim is an animation tool in NS-3. It displays network topology, packet exchange between nodes, node movement and other useful statistics. NetAnim exists in all-in-one package. It takes xml as input and we need to specify the name of this xml file in our application program as given below.

std::string animFile = "sample-animation.xml";

Recommended steps for generating xml animation traces are:

1. Ensure that wscript includes the netanim module as in: src/netanim/examples/wscript.
2. Include the header #include "ns3/netanim-module.h" in the test program.
3. Add the statement AnimationInterface anim ("sample-animation.xml"); before Simulator::Run().

NetAnim is useful for both wired links and wireless links. Figure 7.14 shows the packet animation in a wireless link.

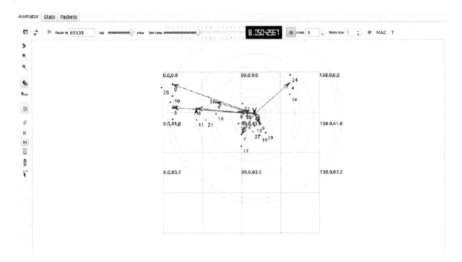

FIGURE 7.14 (SEE COLOR INSERT)
Packet animation in a wireless link

7.6 Summary

The focus of this chapter is to provide a practical guide for programming sensor motes to develop different protocols and algorithms. Developing programs for WSNs in popular platforms and simulators are emphasized in this chapter. The chapter also discusses installation of platforms and simulators. Mainly, TinyOS and Contiki operating systems are considered. In addition, simulators like TOSSIM, COOJA, Castalia and NS-3 are also discussed. TOSSIM is specifically designed for simulating WSN applications for TinyOS platforms. COOJA is a simulator for Contiki platforms. Implementation details of CTP and M-SPIN in TinyOS will help readers understand the use of modules and features of TinyOS. CTP is the most widely used data collection protocol. Broadcast example in Contiki is discussed and TMAC protocol implementation in Castalia is also shown in this chapter. Adding a new mobility model in NS-3 is also discussed in detail.

Bibliography

[1] Castalia download. http://castalia.research.nicta.com.au/index.php/en/download.

[2] Osterlind F., Dunkels A., Eriksson J., Finne N., and Voigt T. Cross-level sensor network simulation with COOJA. In *Proceedings of First IEEE International Workshop on Practical Issues in Building Sensor Network Applications*, 2006.

[3] Castalia installation. `http://castalia.npc.nicta.com.au/pdfs/Castalia\%20-\%20Installation.pdf`.

[4] Seada K., Zuniga M., Helmy A., and Krishnamachari B. Energy efficient forwarding strategies for geographic routing in wireless sensor networks. In *Proceedings of Second International Conference on Embedded Networked Sensor Systems*, 2004.

[5] Academic Public License. `http://castalia.npc.nicta.com.au/pdfs/Castalia\_AcademicPublicLicence.pdf`.

[6] Information NICTA and Communications Technology (ICT) Research Centre of Australia. `http://castalia.research.nicta.com.au/index.php`.

[7] NS-3 manual. `http://www.nsnam.org/docs/release/3.21/manual/ns-3-manual.pdf`.

[8] NS-3 tutorial. `http://www.nsnam.org/docs/release/3.21/tutorial/ns-3-tutorial.pdf`.

[9] Omnet. `http://www.omnetpp.org/documentation`.

[10] Levis P., Madden S., Polastre J., Szewczyk R., Whitehouse K., Woo A., Gay D., Hill J., Welsh M., Brewer E., and Culler D. TinyOS: an operating system for sensor networks. In *Ambient Intelligence*, Springer, 2005.

[11] Bhatti S., Carlson J., Dai H., Deng J., Rose J., Sheth A., Shucker B., Gruenwald C., Torgerson A., and Han R. MANTIS OS: an embedded multithreaded operating system for wireless microsensor platforms. *Mobile Networks and Applications*, 10(4):563–579, 2005.

[12] Dam T. van and Langendoen K. An adaptive energy-efficient MAC protocol for wireless sensor networks. In *Proceedings of First Conference on Embedded Networked Sensor Systems (SenSys 2003)*, 2003.

[13] TinyOS. `http://tinyos.stanford.edu/tinyos-wiki/index.php/Getting\_Started\_with\_TinyOS`.

Index

Printed and bound by CPI Group (UK) Ltd, Croydon, CR0 4YY

24/10/2024

01778302-0006